OPEN-KITCHEN RESTAURANT
开放式餐厅

(荷) 劳拉·沃尔托/编　常文心/译

辽宁科学技术出版社

Contents 目录

006 **Preface: Design Is Merged**
序言：设计即是融合

012 **Chapter 1: Brief Introduction of Open Kitchens**
第一章：开放式厨房简介

014 *1.1 Definition and Forms of Open Kitchens*
1.1 开放式厨房的界定与形式

014 **1.1.1 Definition**
1.1.1 开放式厨房的界定

015 **1.1.2 Forms**
1.1.2 开放式厨房的形式

016 *1.2 Origin and Development of Open Kitchens*
1.2 开放式厨房的起源与发展

017 **1.2.1 Open Kitchens in Houses**
1.2.1 开放式厨房在家居中的应用

018 **1.2.2 Open Kitchens in Restaurants**
1.2.2 开放式厨房在餐厅中的应用

022 **Chapter 2: Design Principles**
第二章：设计准则

025 *2.1 Designing Process*
2.1 设计流程

028 *2.2 Factors to Be Considered*
2.2 需要考虑的因素

034 **Chapter 3: Indoor Air Climate in Open Kitchen Restaurants**
第三章：开放式餐厅的室内空气环境

036 *3.1 Odour and Noise*
3.1 气味与噪声

036 **3.1.1 Odour and Nuisance**
3.1.1 气味

038 **3.1.2 Noise and Nuisance**
3.1.2 噪声

3.2 Ventilation Systems   041
    3.2 通风系统

3.3 Kitchen Hoods   054
    3.3 厨房排风罩

    3.3.1 Selecting & Sizing Exhaust Hoods   054
        3.3.1 排风罩的选择与尺寸

3.4 Ventilated Ceiling   065
    3.4 通风顶棚

    3.4.1 Open Ceiling   066
        3.4.1 开放式顶棚

    3.4.2 Closed Ceiling   067
        3.4.2 封闭式顶棚

Chapter 4: Case Studies   068
第四章：案例分析

An Open Kitchen Is to Entertain Diners   070
    开放式厨房的作用是取悦就餐者

Muvenpick   078
    莫凡彼餐厅

Restaurant Kaskada   084
    卡斯卡达餐厅

Ellas Melathron Aromata   090
    梅斯拉多兰芳香餐厅

NAMUS Buffet Restaurant   098
    纳姆斯自助餐厅

ZOZOBRA Noodle Bar   104
    祖祖布拉快餐店

Kinugawa   112
    鬼怒川餐厅

Vintaged Restaurant at Hilton Brisbane   118
    布里斯班希尔顿酒店复古餐厅

124 *Lluçanès Restaurant in Barcelona*
巴塞罗那卢卡奈斯餐厅

132 *Hilton Pattaya – Restaurants, Lobby & Bar and Linkage Spaces*
芭堤雅希尔顿酒店餐厅、大堂、酒吧和连接空间

140 *Assaggio Trattoria Italiana*
阿萨吉欧意大利餐厅

144 *Sugarcane*
甘蔗餐厅

150 *Symphony's*
交响乐餐厅

156 *De Gusto*
好味道品鉴餐厅

162 *Karls Kitchen*
卡尔斯小厨

170 *The River Café*
河畔咖啡馆

176 *Bar Brasserie Restaurant Fitch & Shui*
费奇 & 舒伊酒吧餐厅

184 *FRESHCUTT*
新鲜出炉烧烤店

190 *Fusao Restaurant*
房雄餐厅

196 *La Oliva*
奥利瓦餐厅

200 *Grand Hyatt Macau – Restaurant "MEZZA 9"*
澳门君悦酒店梅萨餐厅

206 *barQue*
巴Q餐厅

210 *La Nonna Restaurant*
拉诺那餐厅

216 *Jaffa – Tel Aviv Restaurant*
特拉维夫雅法餐厅

Aoyama Hyotei 222
青山冰帝餐厅

Esquire 226
君子餐厅

Chapter 5: Fire Prevention in Open Kitchens 232
第五章：开放式厨房的防火

5.1 Fireproof Units and Partitions 234
5.1 防火单元及分隔

5.2 Safe Evacuation 234
5.2 安全疏散

5.3 Liquefied Petroleum Gas Pot Storage 235
5.3 液化石油气瓶库

5.4 Firefighting Equipment 235
5.4 消防设施

5.4.1 Automatic Fire Alarm System 235
5.4.1 火灾自动报警系统

5.4.2 Automatic Fire Extinguishing System 236
5.4.2 自动灭火系统

5.4.3 Smoke Exhaust System 236
5.4.3 排烟系统

5.4.4 Kitchen Firefighting Equipment 236
5.4.4 厨房灭火设备及器材

5.4.5 Mechanical Air Supply and Exhaust System 237
5.4.5 机械送排风系统

5.4.6 Evacuation Signs and Emergency Lighting 237
5.4.6 疏散指示标志和应急照明

Index 238
索引

Preface:
Design Is Merged

序言：设计即是融合

Preface: Design Is Merged

Design Is Merged

The trend of integrating the kitchen into the restaurant and making it a visible element, started in the beginning of the year 2000. A new food culture emerged which was triggered by the new public interest in gastronomy and the need to reconnect with preparing food. This together with the emerging questions regarding the origin of food, made us rethink the restaurant design as we knew it.

Chefs started branding themselves while the story behind the food became essential. A restaurant wasn't just a place anymore where you could satisfy your desire for good food and social interaction; it became a place where you are entertained and involved as a spectator in the process of preparing the food.

La Place, the famous Dutch restaurant concept, was one of the precursors of this trend in The Netherlands. They introduced an accommodating live cooking concept and showed that there was a need for a more flexible way of eating. Influenced by the Asian food culture where you can eat whatever, whenever and wherever, the concept turned out to be a big hit appealing to a large audience.

This trend also touched down in New York in places such as the Gramercy Tavern, an established example of the evolution of the traditional restaurant lay out to todays design. Other than what you might expect, the best seats in this restaurant are the ones at the bar; it's the interaction with the kitchen what makes these seats so popular. They are closest to the kitchen allowing the guest to be part of the cooking experience. It creates a relationship between the chef and his audience, making them feel special.

To us, Amsterdam-based interior design firm D/DOCK, this is a key factor when we design a restaurant; we want to make people feel special and appreciated by considering them as guests and not as customers. By creating an inviting environment where the guest is in charge and

设计即是融合

在餐厅中引入开放式厨房的潮流始于2000年初。大众对美食和烹饪的兴趣引发了全新的美食文化，他们期待参与到备餐过程中。这一潮流随即要求设计师对餐厅设计进行重新思考。

厨师开始塑造自己的形象，而美食背后的故事变得至关重要。餐厅已经不仅是一个享受美食和进行社交的场所，它已成为了提供厨艺表演的娱乐休闲场所。

荷兰著名的餐厅品牌拉普雷斯是荷兰开放式餐厅的先驱者。他们引入了一种兼容并包的现场烹饪理念，并且指出食客们需要更灵活的就餐方式。餐厅受亚洲美食文化影响，认为人们可以随时随地享用各种美食。这一概念深受广大消费者的欢迎。

开放式餐厅的潮流同样席卷了纽约。葛兰姆西酒店就是从传统餐厅布局到开放式设计的典范。毫无疑问，餐厅中最好的座位就在吧台边；与厨房的互动让这些座位炙手可热。宾客们可以靠近厨房，体验烹饪的艺术。开放式厨房在厨师和他的观众之间建立的联系，使观众感到自己与众不同。

对我们（阿姆斯特丹的D/DOCK室内设计公司）来讲，开放式厨房是餐厅设计的关键。我们希望让人们感到自己不是消费者，而是宾客。开放式厨房让宾客们感到舒适自如，还可以与厨师进行交流互动，是餐厅的中心。人们聚集在开放式厨房四周，正如在家中的厨房炉灶旁一样。

序言　设计即是融合

interaction arises between the different users, the open kitchen becomes the island of the restaurant. It is the place where all people gather around, just like your own cooking island in your kitchen back at home.

Fitch & Shui, one of our latest hospitality concepts in Amsterdam, is such a restaurant. Finger food, pizza and other small bites are prepared at the 20-meter long bar and directly served to the guest sitting there. By doing so, the bar becomes the extension of the kitchen making it the social island of the restaurant.

作为我们在阿姆斯特丹的最新餐饮项目，费奇&舒伊酒吧餐厅正是采用了这种理念。20米长的吧台直接为人们提供小点心、比萨和其他零食，使吧台成为了厨房的延伸，形成了餐厅的社交中心。

Photography/摄影：Render Fitch &Shui: D/DOCK

Preface: Design Is Merged

Photography/摄影：Render Fitch &Shui: D/DOCK

Another example that breaks with all traditional rules of restaurant design is Restaurant Vandaag Amsterdam. Paying by the hour, it is not about how much you eat; it is about how long you are staying. The live cooking stations and the variety of different cuisines and products makes this restaurant feel like a market place where you can shop around for food. After having selected all the ingredients, the chef behind the cooking station will prepare your dish on the spot. This is where the magic happens by turning cooking into entertainment.

另一个打破了传统餐厅设计的是阿姆斯特丹的今日餐厅。餐厅计时收费，并不在乎宾客吃多吃少，在乎的是宾客的停留时间。现场烹饪表演和各式美食让餐厅看起来像是一个自由采购美食的市场。在选定了食材之后，烹饪台后方的厨师就可以现场准备宾客的菜品。这正是将烹饪转化为娱乐的奇妙之处。

Photography/摄影：Alan Jensen

序言　设计即是融合

Photography/摄影：Foppe Peter Schut

The ultimate food experience is the cooking school we designed for BSH Huishoudapparaten at the Amsterdam area, the headquarters of home appliance brands Bosch, Siemens, Gaggenau, Neff and Constructa. Being guided by a professional chef you prepare your own dish and eat it afterwards, which transforms the kitchen into your own restaurant. You become part of the environment by being the chef and the guest at the same time.

To D/DOCK, a good restaurant design is all about creating the ultimate experience. The kitchen, the beating heart of a restaurant, is the primary tool to do so. Creating an open kitchen connects the most important element of a restaurant with its audience, making a restaurant alive and transparent.

D/DOCK
NieuweSpiegelstraat 36
1017 DG Amsterdam
The Netherlands

www.ddock.com
Twitter @DDockDesign
Facebook D/DOCK

我们在阿姆斯特丹的家用电器中心（博世、西门子、嘉格纳、内夫和康斯塔克塔等知名品牌在此都设有总部）所设计的BSH烹饪学校能够提供最顶级的美食体验。专业厨师会指导宾客自己制作美食。厨房摇身一变，成了宾客自己的餐厅，让他们也当了一次大厨。

对D/DOCK设计公司来讲，餐厅设计的秘诀在于创造顶级体验。作为餐厅的核心元素，厨房是最主要的设计手段。开放式厨房能够让餐厅和宾客连接起来，使厨房鲜活而透明。

D/DOCK室内设计公司
荷兰，阿姆斯特丹，1017区，新斯皮格尔街36号

www.ddock.com
Twitter @DDockDesign
Facebook D/DOCK

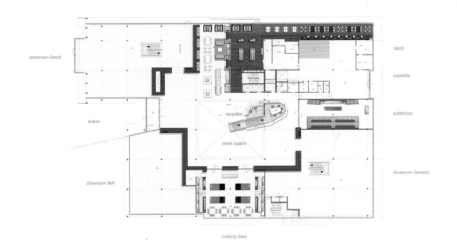
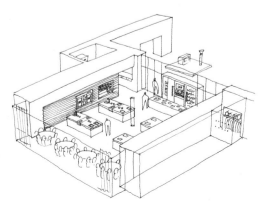

Chapter 1 : Brief Introduction of Open Kitchens

第一章：开放式厨房简介

1.1 Definition and Forms of Open Kitchens

1.1 开放式厨房的界定与形式

1.1.1 Definition

1.1.1 开放式厨房的界定

1.1.2 Forms

1.1.2 开放式厨房的形式

1.2 Origin and Development of Open Kitchens

1.2 开放式厨房的起源与发展

1.2.1 Open Kitchens in Houses

1.2.1 开放式厨房在家居中的应用

1.2.2 Open Kitchens in Restaurants

1.2.2 开放式厨房在餐厅中的应用

Chapter 1: Brief Introduction of Open Kitchens

As interior design goes, the open kitchen is a relative youngster. It is a modern idea, and with it come challenges. In the past, restaurant customers may have preferred food to magically appear out from behind closed doors, with no indication whatsoever about how the sausage is made – figuratively or literally. After years of hearing Big Food and fast food horror stories that'll turn stomach, however, the prototypical modern diner seems to want transparency rather than mystery. (See Figure 1.1, 1.2)

1.1 Definition and Forms of Open Kitchens

1.1.1 Definition
In home space design, an open kitchen refers to the open cooking-dining space that integrates dining table and kitchen through skillful use of space.

In commercial space, an open kitchen refers to commercial kitchen in public space which takes pipelined gas, natural gas or bottled gas as

随着室内设计的发展，开放式厨房仍是一个相对较新的概念。它先进时尚，也面临重重挑战。过去，餐厅消费者可能喜欢看到食物从紧闭的大门内像变魔术一样出现在自己的面前，完全不关心它们的制作过程。但是，在听多了有关食品和餐饮业的不良传言之后，现代就餐者更希望看到整个备餐过程透明化。（见图1.1、1.2）

1.1 开放式厨房的界定与形式

1.1.1 开放式厨房的界定
在家居空间设计中，开放式厨房是指巧妙利用空间，将实用美观的餐桌与厨房紧密相连，形成一个开放式的烹饪就餐空间。

在商业空间内，开放式厨房是指在公共场所，以管

Figure 1.1, 1.2: Open kitchen in the restaurant to show the cooking process in front of diners
图1.1、1.2：开放式厨房将烹饪过程展现在食客眼前

第一章　开放式厨房简介

fuel and non-load transparent component such as tempered glass as partitions or without any partitions to dining area.

1.1.2 Forms (See Figure 1.3, 1.4, 1.5, 1.6)

According to operation requirements, open kitchens can be divided into two forms: full-open kitchen and semi-open kitchen. Full-open kitchen joins kitchen and dining area together to form a whole space. Some restaurants also keep the cooking area closed and preparation area open. The openings of connections between kitchen and other areas vary in size according to design requirements. The opening for pass-through is smaller while the opening for viewing is larger without partitions or fireproof cut-off. According to different show areas, full-open kitchens try to show cooking, washing and sterilisation processes show the diners fully. The kitchen is combined with the front dining area to delete pass-through step and reduce the dish-passing time from kitchen to diners. Meanwhile, the supervision of diners reinforces their interactions. As an

道煤气、天然气或瓶装液化气作为燃料，采用钢化玻璃等非承重透明构件作为厨房与其他部位之间的一般隔断，或者不设任何隔断的厨房与就餐区直接相连的商业用开放式厨房。

1.1.2 开放式厨房的形式（见图1.3、1.4、1.5、1.6）
开放式厨房按经营需要，分为全开放式和半开放式。全开放式厨房是指厨房与就餐区连为一体，共处同一空间。也有的餐饮场所局部将炒菜区封闭、备菜区开放，厨房与其他部位的连接处根据设计的需要开口尺寸大小不一，仅用于传菜功能的连通部位开口就小，而具有观赏性的开口部位就大，没有互相分区更没有防火隔断。全开放式厨房根据能够展现的面积，尽量把烹饪过程、洗碗消毒等环节全方位展示给就餐者。后厨与前店合并在一起，减少了传

Figure 1.3,1.4,1.5,1.6: Open kitchens in restaurants in different forms
图1.3、1.4、1.5、1.6:餐厅中不同样式的开放式厨房

Chapter 1: Brief Introduction of Open Kitchens

innovative management and operation mode, full-open kitchen proposes higher requirement in planning and design.

Semi-open kitchen takes whole-face or large area of non-load transparent material as partition, which means the division material between kitchen and restaurant is not fireproofing. As an example of semi-open kitchen, cooking station uses transparent partition (often a sheet of glass) to connect kitchen and restaurant. The rough processing and cooked food are finished in back of house. Only the food and final step of cooking are shown in front of diners. The diners could enjoy the cooking show freely. This semi-open mode assures diners in the term of sanitation. However, the diners cannot smell the dishes because of the glass partition, which also widens the distance between kitchen and diners. With simple kitchen equipment, semi-open kitchen requires a spacious, bright and aesthetic design. The compounding equipments include show counters and showcases.

1.2 Origin and Development of Open Kitchens

Starting in the 1980s, the perfection of the extractor hood allowed an open kitchen again, integrated more or less with the living room without causing the whole apartment or house to smell. Before that, only a few earlier experiments, typically in newly built upper-middle-class family homes, had open kitchens. Examples are Frank Lloyd Wright's House Willey (1934) and House Jacobs (1936). Both had open kitchens, with high ceilings (up to the roof) and were aired by skylights. The extractor hood made it possible to build open kitchens in apartments, too, where both high ceilings and skylights were not possible.

The re-integration of the kitchen and the living area went hand in hand with a change in the perception of cooking: increasingly, cooking was seen as a creative and sometimes social act instead of work. And there was a rejection by younger home-owners of the standard suburban model of separate kitchens and dining rooms found in most 1900-1950 houses. Many families also appreciated the trend towards open kitchens, as it

菜环节，缩短了菜肴到达就餐者的时间。同时，厨房工作人员在就餐者的监督下工作，增强了他们之间的互动。作为一种创新的管理模式及经营模式，其对厨房的规划设计有了更高的要求。半开放式厨房是指厨房与其他部位用整面或大量使用非承重透明材料作为隔墙，也就是厨房与餐厅之间的分隔材料不具备防火性能。明档即为一种半开放式厨房，在厨房与餐厅衔接位置设置明档（往往是一层薄薄的玻璃），部分粗加工和熟制品在后厨内完成，在顾客面前只展示食品和最后的出餐工艺，厨师直接进行烹调操作，让客人放心享受。这样的开放式，不可否认固然在卫生上产生让人放心的感觉，但遗憾的是顾客的嗅觉完全得不到满足，一道薄薄的玻璃其实在很大程度拉大了厨房与食客之间的距离。半开放式厨房设备简单，规划设计要求宽敞、明亮、美观。配合操作的还有展台及展示柜。

1.2 开放式厨房的起源与发展

20世纪80年代起，吸油烟机的完善为开放式厨房的实现提供了可能，使其可以与客厅相结合，同时又不会使整间住宅充满油烟。在此之前，仅有少数新建的中上层阶级的住宅试验性地采用了开放式厨房。例如，弗兰克·劳埃德·怀特的威利住宅（1934）和雅各布斯住宅（1936）的开放式厨房都采用了高天花板（直达屋顶）并以天窗通风。吸油烟机出现让没有高天花板和天窗的公寓也可以采用开放式厨房。

厨房与客厅的结合改变了人们对烹饪的理解：越来越多的人将烹饪看做创意的社交艺术，而不是工作。1900年–1950年之间，许多年轻的户主拒绝了城

made it easier for the parents to supervise the children while cooking and clear up spills. The enhanced status of cooking also made the kitchen a prestige object for showing off one's wealth or cooking professionalism. Some architects have capitalised on this "object" aspect of the kitchen by designing freestanding "kitchen objects".

Another reason for the trend back to open kitchens (and a foundation of the "kitchen object" philosophy) is changes in how food is prepared. Whereas prior to the 1950s most cooking started out with raw ingredients and a meal had to be prepared from scratch, the advent of frozen meals and pre-prepared convenience food changed the cooking habits of many people, who consequently used the kitchen less and less. For others, who followed the "cooking as a social act" trend, the open kitchen had the advantage that they could be with their guests while cooking, and for the "creative cooks" it might even become a stage for their cooking performance. The "Trophy Kitchen" is highly equipped with very expensive and sophisticated appliances which are used primarily to impress visitors and to project social status, rather than for actual cooking.

Street hawkers in Southeast Asia were the first breed of show kitchens. This was more of a necessity rather than for "show" sake. Japanese teppan-yaki outlets have been utilising this concept for decades. Now, show kitchens are sprouting all over hotels and restaurants in Asia Pacific. In Hong Kong, this uniquely modern phenomenon of "Open Kitchen" concept has been gaining popularity. The kitchen, once literally and figuratively the "back of house", has stepped into the forefront. Restaurateurs are becoming quick to allow roaming eyes into "back of house" areas. The kitchen has taken on a prominent role in providing entertainment, leading the way for "Show Kitchens" and "Open Kitchens".

1.2.1 Open Kitchens in Houses

A simple and spacious open kitchen design will be inevitably favourable to modern people when choosing commercial housing or selecting home decoration. Although with troubles of cooking fumes, many urbanites

郊式的独立厨房和餐厅。许多家庭十分欢迎开放式厨房，因为家长可以边烹饪、清理厨房，边监护儿童。烹饪方式的改进还让厨房成为了炫耀财富或厨艺的工具。一些建筑师以此为出发点，设计了独立的"厨房目标"。

开放式厨房回归的另一个原因（也是"厨房目标"的基础）是食物准备流程的变化。20世纪50年代之前，大多数烹饪都由原材料开始，就餐的准备过程相当长，而冷冻食品和便利食品的出现改变了许多人的烹饪习惯，他们使用厨房的次数越来越少。对于追随"烹饪是社交艺术"概念的人们来说，开放式厨房让他们在烹饪过程中也可以与客人共处，而"创意烹饪"则让厨房成为了厨艺表演的舞台。"身份厨房"配有诸多昂贵精致的设备，主要是为了向客人展示，彰显主人的社会地位，而不是为了实际烹饪需求。

东南亚的街头小贩是最早使用开放式厨房的人群。相对于展示，这主要是出于必要性。数十年来，日本铁板烧摊贩一直采用这种形式。现在，开放式厨房在亚太地区的酒店和餐厅正在兴起。在香港，这种独特的现代"开放式厨房"理念正日渐流行。原来仅指"后厨"的厨房如今已经走到了人前。餐厅迫不及待地向人们展示着后厨区域。在展示厨房和开放式厨房中，厨房已经起到了主要的娱乐作用。

1.2.1 开放式厨房在家居中的应用

一款简洁宽敞的开放式厨房设计，想来在现代人选购商品房或挑选家装风格时都必然会对其青睐有加。虽然顶着油烟四起的烦恼，许多追求家居舒适感与时尚感兼备的都市人依然难以释怀对开放式厨

Chapter 1: Brief Introduction of Open Kitchens

who pursue comfortable sensations and fashion still look forward to open kitchens.

Open plan living spaces have become very popular in recent years as domestic dwellings have become smaller. Creating one large room out of 2 or 3 smaller rooms makes the living space feel larger and more contemporary.

This is an undertaking most often taken up by those whose kids have grown up and left home, or those with disposable income enough to sustain this kind of a change in the layout of their home. Having an open plan kitchen gives your home a flow it otherwise wouldn't have, with room enough for walking space as well as living space and a feel of openness and easy-access to the kitchen area.

A sense of togetherness is achieved, with the ordinarily separate rooms in the house consolidated into a more centralised living space, making the home less compartmentalised. This has its obvious advantages.

The kitchen as wide-open continuation of the living space (rather than tucked-away scullery) brings its own set of questions, such as clear and sensible spatial geometry, definition of function, sufficient wall space for big appliances, and lighting. "The hardest part is visibility. It becomes more than just working space, because everything is on stage and exposed," says Tom Shafer, a principal of Grunsfeld Shafer Architects in Chicago. For a couple who wanted to stay put in the 1950s ranch house they had come to love over many years, Shafer did an extensive remodel, designing the entire house – including the open kitchen – by following the home's existing lines. "They wanted to keep their memories and use the house with the next generation, their married children and grandchildren," Shafer says.

1.2.2 Open Kitchens in Restaurants (See Figure 1.7, 1.8, 1.9, 1.10)
Nowadays, open kitchens are not only popular in houses, many upscale western restaurants and star-rating hotels also arrange them in dining

房的向往。

由于住宅越来越小，开放式布局生活空间在近年十分流行。将两到三个小房间融为一个大房间会使生活空间感到更大、更现代。

开放式布局通常适合孩子已经长大或有可支配收入对住宅进行改造的家庭。开放式厨房能让家居空间更流畅，让房间适合走动和居住，同时还让厨房区域开阔且便于进入。

开放式厨房实现了凝聚感，将住宅中独立的房间合并成一个更集中的生活空间，减轻了家居的分散感。这是显而易见的优势。

延续了生活空间的开放式厨房（而非隐蔽起来的洗碗间）也有自己的问题，例如清晰的空间布局、功能区的划分、安装大型电器的墙面空间以及照明。芝加哥格朗斯菲尔德·沙佛建筑事务所的负责人汤姆·沙佛称："最难的部分是可见性。厨房不仅是工作区，所有东西都被展示出来。"沙佛对一座建于20世纪50年代的乡间别墅进行了改造，沿着住宅的原有线条进行的重建（其中包含开放式厨房）。沙佛说："屋主夫妇希望留住自己的回忆，并且与下一代（他们的子孙）共同使用住宅。"

1.2.2 开放式厨房在餐厅中的应用（见图1.7、1.8、1.9、1.10）
如今，开放式厨房的身影已不仅仅活跃于百姓之家，许多高档的西餐厅和星级酒店也开始将厨房置于客人的就餐环境之中，力求营造一种开放式烹饪的概念和感觉。开放式厨房在餐饮场所的广泛应用已经演变为势不可挡的流行趋势，以现代、时尚、美观、

area to create an open cooking concept and sense. The extensive use of open kitchens has become an overwhelming catering trend. Open kitchens are blended into upscale culinary culture in a modern, fashion, aesthetic and friendly posture and play a significant role.

For diners, open kitchens enable them to view the flow of work of kitchens and unique techniques from chefs. A kitchens is divided into two areas: closed kitchen and open kitchen. The closed kitchen is mainly used for rough processing and chopping and assembling of materials for barbecue dishes and cooking of other dishes, while the open kitchen is responsible of final cooking and final plate of barbecue dishes. This makes cooking a scene feeling. The popularity of cooking show attributes to open kitchen's natures of freedom, freshness and quick upgrade.

For maximum transparency, restaurants ranging from fast-casual superstar Chipotle, to indie eateries favored by foodies, to massive fast-food chains like Domino's are all turning to the open kitchen.

The open kitchen trend seems to have been born in big cities such as New York, where chefs cooked within view of diners largely due to space constraints. Getting in the habit of watching chefs do their thing on TV has obviously boosted the fascination with what goes on in restaurant kitchens. As diners grew obsessed with celebrity chefs and the creative ways fresh and exotic ingredients were being combined, consumers increasingly came to view the flames and steam and clattering in the kitchen as part of the "show" of dining out.

The open kitchen trend trickled down to the growing fast casual market, most obviously with the wildly successful Chipotle chain. The "Chipotle Experience," as it's called, is deeply rooted in transparency, with all the

Figure 1.7,1.8,1.9,1.10: Chefs work in the open kitchen
图1.7、1.8、1.9、1.10:大厨在厨房中忙碌

亲切的姿态融入到高端的餐饮文化中，发挥着不容忽视的作用。

开放式厨房对于就餐者来说，他们可以看到厨房工作流程、厨师的一些绝活表演和特殊技法的演示。厨房被划分为两大不同类型的区域：封闭式厨房区和开放式厨房区。封闭区主要用于烧烤类菜肴原料的粗加工、切配和其他菜肴的制作，而在开放式厨房区域则从事烧烤类菜肴的最后一步烹制和切配装盘。这种使烹饪更具现场性、观赏性的概念能够得到普遍的流行，可以归功于此种形式随意、自然、更新快等与生俱来的优点。

为了实现透明度的最大化，从简餐巨星Chipotle、美食家钟情的私房餐馆到Domino这样的大型连锁快餐都转向了开放式厨房。

开放式厨房的潮流似乎起源于纽约等大城市，一些厨师由于空间所限在就餐者的面前进行烹饪。在电

Chapter 1: Brief Introduction of Open Kitchens

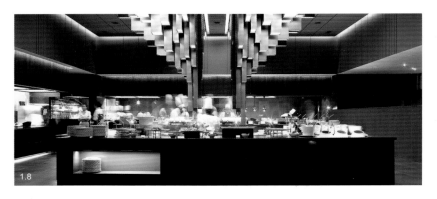 

ingredients of burritos, bowls, tacos, and salads "laid out in front so one can choose the perfect combination to make perfect meal." All of the chopping and assembling of food takes place in full view of customers (behind glass), purposefully so. Here's part of Chipotle's explanation for why the open kitchen design is so important: the sounds, the smells, and the sights of cooking can really help work up an appetite. Unfortunately in a lot of restaurants the "cooking" they do is more like putting together a science experiment. To that end, each Chipotle is designed with a kitchen that's open to the entire restaurant.

QSR Magazine named "Transparency" one of the biggest quick service restaurant trends, anticipating that more restaurants will "follow the trend of open kitchens as a sign to customers that they have nothing to hide." By now, the open kitchen has spread to smaller cities such as Milwaukee. "I do think Milwaukee is catching up to a more national trend," said one chef in the city. "Thanks to celebrity chefs and good food, the dining public wants to see what's going on. Also chefs, me included, are proud of what we do and like to showcase our habitat."

"Open Kitchen of Flame Cooking" at Towngas Avenue (See Figure 1.11)
"Open Kitchen of Flame Cooking" in Towngas Avenue is more than the

视上观看厨师烹饪也激发了人们对餐厅厨房的好奇心。随着就餐者越来越迷恋名厨和新鲜食材与异域食材的混合方式，越来越多的消费者到餐厅里体验厨房的火焰、蒸汽和炒菜的哗啦声，并将其作为了"外出就餐秀"的一部分。

开放式厨房的潮流发展到了简餐市场，最明显的就是Chipotle连锁店的成功。"Chipotle体验"深深植根于透明度，所有制作墨西哥卷、炸玉米饼和沙拉的材料都摆放在就餐者面前，人们可以选择完美的组合来实现完美的一餐。所有食物切配工作都在消费者面前（隔着玻璃）呈现。以下是Chipotle对开放式厨房设计的重要性的解释：烹饪的声音、味道和画面有助于引起食欲。遗憾的是，许多餐厅的烹饪过程就像科学实验一样。因此，每间Chipotle餐厅都配有一个向整个餐厅开放的厨房。

《QRS》杂志将"透明度"列入了快速服务餐厅的重要潮流之一，并预期越来越多的餐厅将"跟随开放式厨房的潮流，向消费者表示他们毫无隐瞒"。

第一章　开放式厨房简介

expansion of the kitchen function. It performs a dual function of production and entertainment. Not only they can enjoy different cuisine, premium beverages and freshly baked snacks prepared by "Flame Cooking", but also enjoy the entertainment produced by "Flame cooking". The "Show Kitchen" provides suspense and action for the diners, through the flashes of fire, the sounds of food sizzling, and the chef's chopping and cooking skills. Diners can have a unique dining-out experience provided by this "Open Kitchen of Flame Cooking".

Generally speaking, the open kitchen is still a concept. The degree of opening needs to be decided by actual requirements. Of course, people could also achieve open kitchen concept in their unique methods.

1.10

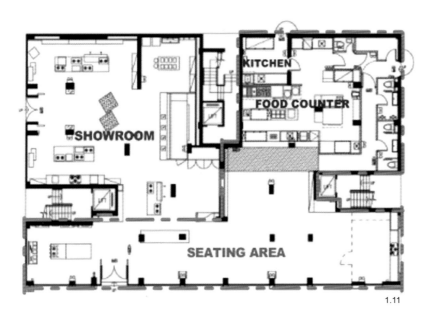
1.11

目前，开放式厨房已经发展到了密尔沃基这样的小城市。城中的一位厨师称："密尔沃基在这方面已经赶上了全国的潮流。感谢名厨和美食让就餐者想要看到烹饪过程。我们厨师都感到骄傲，像是在展示自己的家一样。"

唐加斯大道"火焰烹饪开放式厨房"（见图1.11）位于唐加斯大道的"火焰烹饪开放式厨房"项目不仅是对厨房功能的扩展，它更具有生产和娱乐双重功能。人们不仅可以享用"火焰烹饪"所提供的不同的美食、顶级的酒水和新鲜出炉的点心，还能欣赏他们的表演。"展示厨房"通过火光、食物的滋滋声和厨师的刀工和烹饪技巧为就餐者制造了悬念和动感。就餐者在此可以得到独特的外出就餐体验。

总体来讲，开放式厨房其实还是一种观念，可以根据实际需要决定开放程度，当然更可以通过自己独特的方法来实现开放厨房的理念。

Figure 1.11: Layout plan of Towngas Avenue in Causeway Bay
图1.11：唐加斯大道餐厅平面布局

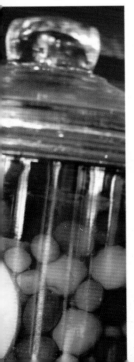

Chapter 2 :
Design Principles

第二章：设计准则

2.1 Designing Process

2.1 设计流程

2.2 Factors to Be Considered

2.2 需要考虑的因素

Chapter 2: Design Principles

"Open kitchens (display kitchen, show kitchen or presentation kitchen) are rapidly gaining prominence, not only because they offer a way to express the concept of a restaurant, but they also attract guests' interest towards the process of cooking", Praveen K Singh once pointed out.

Often the idea of an open kitchen or display kitchen is categorised as a trend in kitchen design. In fact, it is a trend in restaurant concepts. And by "restaurants", it means almost every location where customers gather to select and enjoy food; therefore traditional restaurants, food courts, canteens at institutions and healthcare dining facilities for ambulatory patients, staff, visitors and guests are all included in the definition.

Says M Ram Vittal Rao, hotel consultant, who has designed the largest number of interactive display kitchens in the country, "I can certainly say that an enormous proportion of the kitchens we craft today include an open or display area. In most cases, a major part of the kitchen is in view of ordering and dining customers. 60 to 80 percent of the food which is in the process of preparation or at least in the final stages occurs in full view of customers, so as to reinforce freshness and to enhance customers' appetites."

The design of an open kitchen is invovled with several principles and amount of factors need to be considered. (See Figure 2.1, 2.2)

普拉维恩·K·辛格指出："开放式厨房（又称展示厨房）正快速地成为潮流趋势，不仅是因为它们展示了餐厅的理念，还因为它们吸引顾客注意烹饪的流程。"

开放式厨房或展示厨房的概念通常被归类为厨房设计的趋势。事实上，它是一种餐厅理念的趋势。这里的"餐厅"指的是消费者聚集起来选择和品尝美食的场所，因此，传统餐厅、美食广场、学院机构的食堂、医疗机构的餐饮设施等都包含在这个概念之内。

拥有丰富的互动式展示厨房设计经验的酒店顾问M·拉姆·维塔尔·拉奥称："我可以肯定现在许多厨房都包含一个开放区或展示区。很大一部分厨房都向消费者进行了展示。60%~80%的备餐流程或最后的出餐过程都展示在消费者面前，增加了新鲜感并促进了食欲。"

开放式厨房的设计涉及一系列设计准则和需要考虑的因素。（见图2.1、2.2）

Figure 2.1,2.2: Food are prepared in front of diners
图2.1、2.2：在食客面前展示食物烹饪过程

2.1 Designing Process

It remains a fact that the design process has an established sequence of events that must be completed for the best results. Each stage in the process builds on work completed in the previous stages – it is a cumulative effort.

Master planning is the first step in the process, where one establishes what the project should consist of, or if it should exist at all. The main objectives of this phase are justification of the project and exploration of potential design solutions

Programming
Once one has determined that the project is at least worth exploring, a narrative should be developed that will describe the scope of the project as well as its various "components", such as the individual areas within a single facility, or facilities that comprise a greater complex. This narrative will often include required adjacencies (areas that must be located next to each other), desired sizing, and any special requirements or considerations. The programme communicates the owner's desires and requirements to the design team in written form.

Concept Design
This is often the first time that the effort includes any form of graphic communication. The conceptual design is the first attempt to translate the written programme into graphic form. It will typically consist of "bubble diagrams", which identify the anticipated location and size of each area in the program. It may also consist of renderings that highlight the exterior, entry, or other key design elements of the project.

Schematic Design
This phase in the design process builds on the concept phase through the development of block/schematic drawings, geared to identify elements such as walls, counters, and key pieces of furniture or equipment. These drawings focus on defining the footprint of the space, but do not contain a

2.1 设计流程

设计流程是一系列既定事件的序列。设计流程的每个阶段都建立在前一阶段完成的基础上，是一种累积效果。

总体规划是流程的第一步，确定了项目的组成部分或可行性。这一阶段的主要目标是判断项目是否可行并探索潜在的设计方案。

规划
一旦确定了项目的可行性，下一步就是对项目的范围及其组成元素进行描述，例如是独立设施内的独立区域还是大型综合体内的某个设施。描述通常包括要求的邻接限制（必须相邻的区域）、预期尺寸以及其他特殊要求或考量。项目规划以书面形式向设计团队表达了业主的期望和要求。

概念设计
概念设计首次涉及图形表达形式，是将书面规划转换为图形形式的第一步。它通常由"气泡图"组成，明确规划中每个区域的预期位置和尺寸。概念设计可能还包括强调外观、入口或其他重要设计元素的效果图。

方案设计
这个设计阶段建立在概念阶段的基础上，通过体块图/示意图来确定墙壁、台面、主要家具或设备等元素。这些图纸聚焦于确定空间印记，而不包含体块内的细节。

Chapter 2: Design Principles

great deal of detail within these "blocks".

Design Development
At the end of this stage, the general floor plan is typically locked in and the detail within each space must be defined and developed. While a restaurant's locations and dimensions might have been identified in the Schematic Phase, it is in Design Development where the individual pieces of equipment are selected and incorporated into the design. At the end of this phase, it should be expected that all walls, furniture, fixtures, and equipment will have been included in the drawings and clearly identified. While it is not necessary to know the manufacturer of an individual piece of equipment in the restaurant, it is important to know what equipment is required, as well as the required dimensions and configuration.

Construction Documents
Also called "working drawings", this stage consists primarily of the mechanical, electrical, and plumbing (MEP) coordination required to make the building function. The "systems" within a building are very similar to those which exist in the body when you think about it – Structural (skeleton), Mechanical (breathing), Electrical (basis of cell communication and activity), plumbing, etc. This phase requires a significant amount of coordination between a large number of disciplines. It is an extremely important aspect of the design process, as a mistake at this stage can be very costly.

Specifications
Once the construction documents are completed, written specifications are developed to convey all information to the construction team. These specifications include details on the manufacturer, model number, and any required options for everything from the door hardware to the paint, and the flooring materials to the foodservice equipment. These specifications are then used by the construction team to gather pricing from general contractors and their sub-contractors. (See Figure 2.3, 2.4, 2.5, 2.6)

设计开发
在这一阶段的最后，通常会决定总平面图和各个空间内的细节。餐厅的位置和规模可能在方案设计阶段已经确定下来，而设计开发阶段则主要对独立设备进行选择，并将其融入设计。在这一阶段的末尾，需要在图纸中明确所有墙壁、家具、固定装置和设备的位置。尽管没有必要明确餐厅内各个设备的制造商，但是要了解所需设备的种类、尺寸和配置。

施工文件
施工文件又名"施工图"，这一阶段主要与机械、电气和管道等功能设施相关。建筑中的各个系统与人体系统相似：结构系统（骨架）、机械系统（呼吸）、电气系统（细胞通讯和活动的基础）等。这一阶段要求各学科之间的通力合作，在设计流程中至关重要。任何一个错误都可能造成极大的损失。

性能规范
在完成施工文件之后，必须以书面形式将性能规范传达给施工团队。这些性能规范包括制造商的详细情况、型号以及有关五金、涂料、地板材料、食品服务设备等方方面面的选择要求。施工团队将利用这些规范从总承包商和分包商那里收集报价。（见图2.3、2.4、2.5、2.6）

第二章　设计准则

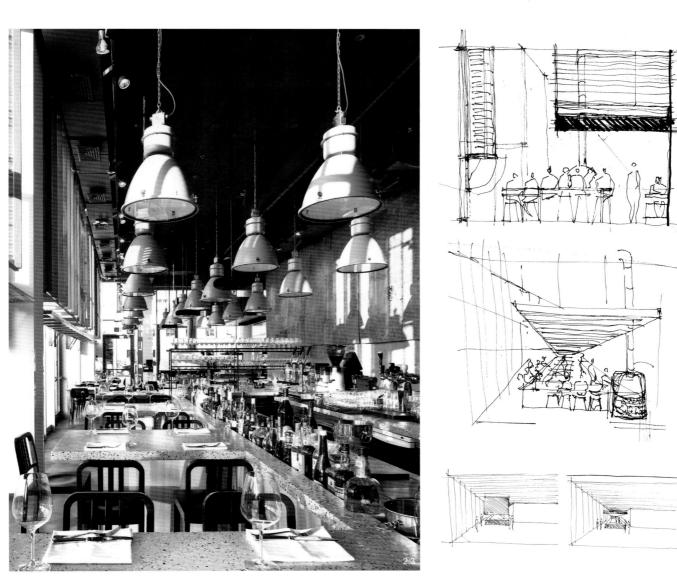

Figure 2.3,2.4,2.5,2.6: Hand sketch and the realised open kitchen
图2.3、2.4、2.5、2.6：手绘稿和最终实景

Chapter 2: Design Principles

2.2 Factors to Be Considered

Space Planning

"Getting into the nuances of space planning, for example, a 40 to 60-seater restaurant will need 800 to 1200 square feet (11.5 square metres) if it is a fast food operation or 1000 to 1500 square feet (139.4 square metres) if it is a fine dining," says Rao.

He also elaborates that, in the case of fine dining, however, the space specified above is devoid of area for the buffet counter for which another 150 square feet (14 square metres) of space would have to be added. Moreover, should the fine dining restaurant have an open kitchen, then 120 to 150 square feet (14 square metres) additional space would be required. Rao continues, "A bar which would just dispense would need about 80 to 100 sqare feet (7.4 to 9.3 square metres) space in addition to the space which has been mentioned above for the restaurant. Therefore in total, a fine dining restaurant of 40 seats with a buffet counter, display kitchen and a bar would amount to an area of approximately 1450 square feet (135 square metres)."

Layers of Activity

The visual impact of an open kitchen is significantly enhanced when there a variety of activity that occurs at different levels of depth perception. In other words, it looks really cool when there is a whole bunch of stuff going on simultaneously in different parts of the open kitchen. On way to achieve this is to incorporate pieces of equipment with constant activity. Rotisseries, open hearth ovens, charbroilers, and even six-burner ranges are good examples of equipment that have constant activity.

This concept can be taken a step further when the stations are strategically located. For instance, picture a display kitchen with a pick-up counter at the front, an expediter's station on the chef's side behind that counter, with an island cooking suite behind that (lots of activity there), and an open hearth oven with a constant flickering flame behind that. This layout has the potential for a variety of activity to occur simultaneously,

2.2 需要考虑的因素

空间规划

拉奥提出："要注意空间规划的差别。例如，同样为40~60座的餐厅，快餐店只需要11.5平方米的空间，而正式餐厅则需要139.4平方米。"

此外，在正式餐厅中，还应在以上基础上增加14平方米的自助吧台空间。如果正式餐厅要设计开放式厨房，那么还需要增加14平方米。拉奥称："酒水吧的设置需要7.4~9.3平方米的附加空间。因此，一家40座的正式餐厅如果配置自助吧台、展示厨房和酒水吧，其总面积应当在135平方米左右。"

活动层次

当各种活动在不同层次进行时，可以大大提升开放式厨房的视觉效果。也就是说，当开放式厨房的各个部分都同时进行活动时，看起来会十分精彩。恒定运转的设备能够实现这种效果，例如烤肉架、平炉烤箱、烧烤炉乃至西餐炉等。

有策略地设置台面能深化这一效果。例如，将带有取菜台的展示厨房设在前方，装盘台设在吧台后方厨师一侧，后面是烹饪区（活动较多），而背景为火光闪烁的烤炉。这种布局可以让多种活动同时进行且互不干扰，这就是"活动层次"概念。（见图2.7、2.8）

one behind the other. This is the concept of "layers of activity". (See Figure 2.7, 2.8)

How's the View?

The objective of an open kitchen is to prepare food in front of the guest. It would make sense, then, that the open kitchen be located so that it can be easily viewed from the dining room. In certain scenarios it is appropriate for the kitchen to be visible from as much of the dining room as possible, while in other situations the kitchen may only be visible from a portion of the dining room. There is no right or wrong answer, rather such a decision depends on the overall concept. The location of the open kitchen, however, is an important issue that can affect the facility's design and should be addressed early in the design process.

视野

开放式厨房的目标是在顾客面前展示备餐过程。因此，开放式厨房的选址必须让就餐区可以轻易地看到。通常来说，应当让厨房展示在更多的就餐空间之前；一些情况下，部分就餐区可能无法看到厨房。设计没有标准答案，一切都取决于整体规划。开放式厨房的位置可以影响餐厅的设计，必须在设计流程早期进行确定。

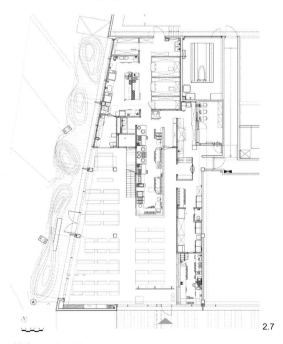
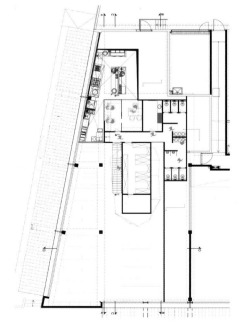

Figure 2.7, 2.8: Layout plan of an open-kitchen restaurant
图2.7、2.8：开放式厨房平面布局

Chapter 2: Design Principles

Flow

The importance of flow has to be addressed within a foodservice facility in depth. Flow is even more important for open kitchens, primarily because the interface between the service staff and kitchen staff occurs in full view of the dining area. Under normal circumstances, this interface can be chaotic at times.

In open kitchens, chaotic conditions simply are not an option. The service staff and kitchen staff need to meet, but never cross paths. The culinary staff should be able to access all support areas and produce their food without interruption from the servers. The servers, likewise, should be able to access all required areas without having to navigate the culinary staff. Because open kitchens are often divided, with some areas in full view and others concealed, achieving proper flow patterns can be more challenging than in traditional configurations. Proper consideration should be given to the final layout. (See Figure 2.9, 2.10)

A Happy Medium

Open kitchens, due to their intricate planning, configuration, and expense, do not typically provide a great deal of flexibility. Concealed, back-of-house kitchens are better suited for equipment configuration changes or substitutions. So, they can limit flexibility despite their higher costs. Fortunately, there is a happy medium when an open kitchen is desired but the cost and limitations are prohibitive.

Instead of having an entire kitchen in full view of the guest, it is possible to incorporate a remote display station within view from the dining area. The idea behind such a station is to offer the ambiance and interactivity unique to open kitchens, while limiting the expense and increasing the flexibility of the facility. In such a configuration, the main kitchen (i.e. hot and cold production, food pick-up areas, and service support stations) would remain in the back-of-house. Only a limited amount of production would be in view of the guest.

流畅性

在食品服务设施中，流畅性十分重要。尤其是开放式厨房，服务员与厨师的接触活动完全暴露在就餐者面前。在正常情况下，场面有时会很混乱。

在开放式厨房中，不能发生混乱的状况。服务员和厨师需要会面，但是他们的路线不能交叉。厨师应当能够进出所有辅助区域，在不受服务员打扰的前提下制作食物。由于开放式厨房通常会进行分区，一些区域会隐藏起来，实现合适的流畅交通模式更具挑战。应当仔细考虑最终的布局。（见图2.9、2.10）

折中办法

由于复杂的规划、配置和较高的花费，开放式厨房的灵活性一般不强。隐藏起来的后厨更适合更换设备或改变配置。所以，尽管造价较高，开放式厨房仍然限制了灵活性。然而，有一个折中办法可以兼顾开放式厨房的功能、造价和限制性。

可以不让整个厨房展现在顾客面前，而是设置一个从就餐区可以看到的展示台。展示台能提供开放式厨房特有的氛围和互动感，同时还能限制成本，增加设施的灵活性。在这种配置中，主厨房（即冷热餐制作、传菜区和服务台）保留在后厨，只有一小部分美食展示在顾客面前。

例如，甜品台就十分适合这种配置。可以安排一到两名厨师配置甜品，还可以用平炉烤箱营造出繁忙的氛围。顾客能看到巧克力酱浇在盘子上的过程，还能看到厨师从烤箱中拿出新鲜烘焙的甜品。想象一下，香气会从甜品台一直弥漫到整个餐厅，多么完美。

Dessert stations, for example, are well-suited for such a configuration. Picture a station with one or two culinary team members assembling desserts. Perhaps there is an open hearth oven for baked desserts to offer the constant activity desired. Guests will be able to see the chocolate sauce generously added to each plate and watch the chefs pull fresh baked desserts form the oven. Imagine the smell that would emulate from this station and drift throughout the dining room. Mission accomplished!

协调

正如我们所知，具有良好规划和设计的开放式厨房能够显著提高就餐体验。开放式厨房的协调要求非常高，因为它们具有独特的形式和功能。要想开放式厨房实现预期的功能，协调是必不可少的。必须严格分析和策划每个细节。经过良好的规划，开放式厨房可以成为视觉的盛宴。

Figure 2.9, 2.10: Diners can enjoy the cooking process during having delicious food
图2.9、2.10：食客可以在享受美食的过程中观赏烹饪过程

Chapter 2: Design Principles

Coordination

As we have seen, open kitchens can significantly enhance the dining experience when properly planned and coordinated. Open kitchens require a tremendous amount of coordination, as they feature unparalleled integration of form and function. If an open kitchen is to achieve its desired function, coordination is critical. Every detail must be analysed and conceptualised. When properly planned, open kitchens can be a feast for the eyes!

美观因素

开放式厨房中的美观因素十分重要。如果设计得当，甚至连辅助设备和配置都能影响视觉效果。储藏区应当隐藏起来或保持整洁。固定设备必须比独立设备更令人满意，以实现完美的效果。美观因素的设置将会影响运营和施工预算。（见图 2.11、2.12）

2.11

第二章　设计准则

Aesthetic Component
The aesthetic component is important in open kitchens. When designed properly, even the support equipment and configuration takes into account what the guest will see, and how it looks. Storage areas will have be concealed or kept neat. Built-in equipment may be more desirable than free-standing equipment so that a finished look is achieved. Both the operation and construction budgets can be impacted by decisions made to accommodate the aesthetic objectives. (See Figure 2.11, 2.12)

Interior Finishing Materials
Due to extensive use of cooking oil and dried food, hazardous sources such as fuels and fuel gas and inflammable things such as furniture, decorations, curtains, tablecloths and carpets, when the open kitchen is connected with other parts of the restaurant, the fire is easy to spread quickly. In consideration of dinning places' specificities, it is recommended to select materials with high fire resistance as interior finishing materials to ensure safety. Besides incombustible materials used in ceiling, walls and floor, the rigorous control of the combustion performance of decorative materials, fabrics and furniture is especially important. The ceiling, walls and floor of dining area should use incombustible or nonflammable materials as possible. The selection of cables and wires, shells of electric equipment, fixed furniture such as tables, sofas and cabinets, carpets, curtains and decorative objects should conform to GB20286-2006 of China.

内部装修材料
由于开放式厨房的餐饮场所，大量使用的食用油、干货食品等，各种燃料、燃气等危险源较多，餐厅的家具、装饰物、窗帘、台布、地毯等可燃物多，在厨房与餐厅等其他部位连通的情况下，火灾时更易加速火灾蔓延。鉴于餐饮场所的特殊性，在进行内部装修选材时，宜采用防火性能高的装修材料，以提升安全性。除厨房内从顶棚、墙面到地面的装饰材料均采用不燃材料外，严格控制餐厅内装饰材料、装饰织物、家具等燃烧性能显得尤为重要。就餐区的吊顶、墙面、地面的装饰尽量采用不燃或难燃材料，控制可燃、易燃装修材料；电线电缆、电器设备外壳、桌椅沙发橱柜等固定家具、地毯、窗帘、装饰物品等应选用符合 GB20286-2006《公共场所阻燃制品及组件燃烧性能要求和标识》标准的产品。

Figure 2.11, 2.12: Delicate decoration of an open-kitchen restaurant
图2.11、2.12：开放式厨房餐厅精美装饰

Chapter 3: Indoor Air Climate in Open Kitchen Restaurants

第三章：
开放式餐厅的室内空气环境

3.1 Odour and Noise

3.1 气味与噪声

3.1.1 Odour and Nuisance

3.1.1 气味

3.1.2 Noise and Nuisance

3.1.2 噪声

3.2 Ventilation Systems

3.2 通风系统

3.4 Kitchen Hoods

3.3 厨房排风罩

3.3.1 Selecting & Sizing Exhaust Hoods

3.3.1 排风罩的选择与尺寸

3.4 Ventilated Ceiling

3.4 通风顶棚

3.4.1 Open Ceiling

3.4.1 开放式顶棚

3.4.2 Closed Ceiling

3.4.2 封闭式顶棚

Chapter 3: Indoor Air Climate in Open Kitchen Restaurants

Today's food service operations are an experience for the senses. Dining in restaurants is more than a meal away from home; it is a carefully orchestrated show: the diners are the audience, the chefs are the stars, and the visible portion of the kitchen is the stage. The display/open kitchen has evolved into entertainment, a carefully choreographed hive of activity just a few feet from the table.

A comfortable indoor environment is a key factor in running a successful foodservice business and providing excellent service. In addition to the efficiency and functionality demands of traditional kitchens, display/open kitchens pose a unique aesthetic and acoustic challenge as well as significant comfort demands for the benefit of guests.

Open kitchens are typically characterised by heavyload appliances located inside the dining area. When heavy cooking equipment is installed (grills, char broilers, satay grill, charcoal grill, Teppanyaki table, teriyaki grill etc...), a significant amount of oduour, smoke or noise can be generated in a very short time. What is most important is that the guests can enjoy the cooking show without being disturbed by cooking emissions or odours. (See Figure 3.1, 3.2)

3.1 Odour and Noise

3.1.1 Odour and Nuisance

Objectionable and offensive odours can cause significant adverse effects on people's lives and well being. Information on the odour nuisance from commercial kitchens is limited although the Chartered Institute of Environment Health (CIEH) has carried out surveys of Local Authorities to quantify the level of complaints made relating to odour nuisance in general. Table 3.1 lists the information gathered by the CIEH during the survey. The survey results:
· do not differentiate between types of industrial process;
· do not specifically identify commercial kitchens; and
· relate to all complaints received and not only "justified" complaints.

It is anticipated that odour problems associated with commercial kitchens will form only a small proportion of the complaints received but will form a significant proportion of the "premises subject to complaint".

当今的食品服务运营越来越注重感官体验。在餐厅就餐已经不仅是外出就餐，而是一场精心打造的演出：就餐者是观众，厨师是明星，开放式厨房则是舞台。开放式厨房已经进化成一种娱乐设施，是距离餐桌几步之遥的热闹场所。

舒适的室内环境是成功的食品服务的关键。除了传统厨房所的效率和功能要求之外，为了让顾客感到舒适，开放式厨房的设计还面临独特的美学和音效挑战。

开放式厨房通常将重型设备放置在就餐区内。重型烹饪设备（烤架、烧烤炉、沙嗲烤架、炭烤架、铁板烧台、照烧烤架等）会在短时间内产生大量的气味、油烟或噪声。最重要的是，要让顾客在不受烹饪油烟或气味干扰的情况下欣赏厨艺表演。（见图3.1、3.2）

3.1 气味与噪声

3.1.1 气味
令人厌恶的气味会对人们的生活和健康造成巨大的负面影响。有关商用厨房气味公害的信息十分有限，但是英国环境卫生协会对地方当局所接到的整体气味公害投诉进行了调查。表3.1列出了英国环境卫生协会的调查信息。调查结果具有以下特点：
- 不区分产业生产的类型
- 不特指商用厨房
- 包含所有投诉，不仅限于合理的投诉

商用厨房的气味问题可能只占所有投诉的一小部分，但是在"经营场所投诉"中会占有很大的比例。

Figure 3.1, 3.2: In the open kitchen, chefs illustrate step by step what they are cooking
图3.1、3.2：厨师们在开放式厨房内演示他们的烹饪步骤

Table 3.1 Results of CIEH survey of odour complaints attributed to industrial processes
表 3.1 英国环境卫生协会有关产业生产气味投诉的调查结果

Year 年	1998/1999	1999/2000
Complaints received 收到的投诉	8,970	10,135
Complaints per million population 每百万人的投诉率	339	358
Premises subject to complaint 经营场所投诉	3,951	5,243
Notices served 整改通知	80	60
Prosecutions 执行	0	5
Convictions 获罪	0	5

Chapter 3: Indoor Air Climate in Open Kitchen Restaurants

Effects of odour
The main concern with odour is its ability to cause an effect that could be considered "objectionable or offensive". An objectionable or offensive effect can occur where an odorous compound is present in very low concentrations, usually far less than the concentration that could cause adverse effects on the physical health of humans or impacts on any other part of the environment.

Effects that have been reported by people include nausea, headaches, retching, difficulty breathing, frustration, annoyance, depression, stress, tearfulness, reduced appetite, being woken in the night and embarrassment in front of visitors. All of these contribute to a reduced quality of life for the individuals who are exposed.

Factors that influence magnitude of an odour problem
Factors that influence the control of odour from commercial kitchens include:
·Size of the cooking facility: This influences the intensity of the odour and volume of ventilation air to be handled.
·Type of food prepared: This affects the chemical constituents within the ventilation air.
·Type of cooking appliances used: This dictates the level of fat, water droplets and temperature within the ventilation air.

3.1.2 Noise and Nuisance
Noise is one of the main environmental problems in Europe, potentially affecting people's health and behaviour. Noise is generated by several types of source such as transport, indoor and outdoor equipment and industrial activity. In the case of commercial kitchens the noise generated by them can affect employees and the surrounding neighbourhood.

Information on the noise nuisance from commercial kitchens is limited. The CIEH have carried out surveys of Local Authorities to quantify the level of complaints made relating to noise nuisance in general. Table 3.2 lists the information for Commercial/Leisure activities for 2002-2003 in England and Wales. The survey results:

气味影响
气味的主要问题在于它会产生"令人厌恶或具有攻击性的"影响。低浓度的气味混合就能造成这种影响。这种浓度可能远小于会对人体健康或环境造成负面影响的浓度。

人们感到的气味影响包括头晕、头疼、恶心、呼吸困难、失望、厌恶、抑郁、压力、流泪、食欲下降、在夜间被熏醒和在访客面前的尴尬。以上影响都能降低受害人的生活品质。

影响气味问题等级的因素
影响商用厨房气味控制的因素包括：
• 烹饪设施的规模：这将影响气味的浓度和需要处理的通风量
• 所准备食物的类型：这将影响通风气流中的化学成分
• 所使用的烹饪设备的类型：这将确定通风气流中的油、水滴和温度的等级

3.1.2 噪声
噪声是困扰欧洲的主要环境问题，能够影响人们的健康和行为。噪声源有几种类型，如交通、室内外设备和产业活动。商用厨房中的噪声可以影响员工和周边的居民。

有关商用厨房气味公害的信息十分有限，但是英国环境卫生协会对地方当局所接到的整体噪声公害投诉进行了调查。表3.2列出了2002年~2003年英格兰和威尔士的商业/休闲活动信息。调查结果具有以下特点：

- do not differentiate between types of commercial/leisure activities;
- do not specifically identify commercial kitchens; and
- relate to all complaints received and not only "justified" complaints.

It is anticipated that noise problems associated with commercial kitchens will form only a small proportion of the complaints received and will also form a small proportion of the "sources complained of".

- 不区分商业/休闲活动的类型
- 不特指商用厨房
- 包含所有投诉，不仅限于合理的投诉

商用厨房的噪声问题在所有投诉和来源投诉中只占很小的比例。

Table 3.2 Results of CIEH survey of noise complaints attributed to commercial/leisure activities in England and Wales (CIEH, 2003)
表 3.2 英国环境卫生协会有关英格兰和威尔士商业/休闲活动噪声投诉的调查结果（CIEH, 2003）

Category of noise nuisance 噪声类型	Commercial/leisure 商业/休闲
Complaints received 收到的投诉	40,602
Complaints per million population 每百万人的投诉率	1,014
Sources complained of 来源投诉	32,302
Sources confirmed as a nuisance 已确认的来源	4,771
Nuisances remedied without notices being served 无整改通知而主动改正的	3,140
Notices served 整改通知	1,310
Prosecutions 执行	95
Convictions 获罪	48
Nuisances remedied by Local Authority in default 当局缺席的整改	125

Chapter 3: Indoor Air Climate in Open Kitchen Restaurants

Types of noise in industrial kitchens

Factors that influence magnitude of noise in a commercial kitchen are:
·Size and format of the exhaust: The bulk flow leaving the exhaust diffuser generates broadband aero-acoustic noise. The sound level increases with increase in air speed and decreases with increase in area. The presence of grilles will generate tonal components. The sound levels are inversely proportional to the increase in area and increase with the eighth power of the flow speed.
·Heat release from kitchen: This influences the size of the exhaust system required and the flow rate of air to be handled by the system. Increase in flow rates can increase the pressure perturbations that can generate noise or can excite other parts of the system leading to noise.
·Type of cooking appliances used: this dictates the overall noise level as each individual appliance might contribute significantly to the total noise.
·Position of exhaust fan in the system: This may influence the noise radiated by the fan to the interior or exterior of the building and the transmission of sound energy into the exhaust duct system.
·Fitting and dimensions of the exhaust flow ducts: exhaust duct dimensions, fixings and insulation can all influence the amount of noise these structures will transmit and propagate. Selection of appropriate noise attenuating materials, avoidance of flow restrictions, and vibration isolators between the ducts and the fan are some of the aspects to be considered.
·Fan type and speed: Type of fan used (e.g. centrifugal fan with blades that are backward curved, forward curved or radial, or axial fan) will influence the level and nature of noise emitted. The fan characteristic needs to be chosen so that it is operating at its most efficient duty point as this tends to be the region of minimum noise. If fan speed is too high it will be operating away from that point which can lead to increases in level of up to 10 dB, as well as inefficient air management. It is often also desirable acoustically to use larger fans operating at low speeds rather than smaller fans operating at higher speeds. (See Table 3.3)

商用厨房中的噪声类型

影响商用厨房噪声等级的因素包括：
• 排气装置的规模和格式：排气扩压器的总体气流会产生宽频空气噪声。噪声等级随着风速的增加而增强，随着面积的增加而减弱。隔栅会产生单音波。噪声等级与面积成反比，以风速的八次方增强
• 厨房释放的热量：这将影响排气系统的规模及其所处理的空气流量。流量的增加会增大压力扰动，从而产生噪声或引起系统的其他部分发出噪声
• 所使用的烹饪设备的类型：这决定了总噪声等级，每件独立的设备都会产生相应的噪声，最终叠加起来
• 系统中排风扇的位置：这会影响排风扇向室内外辐射的噪声以及排气系统中声能量的传递
• 排气管的装配和尺寸：排气管的尺寸、装配和绝缘能够影响这些结构所传递和传播的噪声总量。选择合适的减噪材料、避免节流限制、在管道和风扇之间安装隔震器都是有效的措施
• 排风扇类型和转速：排风扇的类型（如带有后倾式扇叶、前倾式扇叶或辐射式扇叶的离心式风扇、轴流式风扇）将影响噪声的等级和性质。风扇特性的选择十分重要，关系到它的最大效率点和最低噪声领域。如果风扇转速过高，它将脱离最大效率点，可能导致噪声增加10分贝，还会影响空气处理效率。使用大型低速风扇也比小型高速风扇所产生的噪声要少。（见表3.3）

Table 3.3 Sources of noise from commercial kitchen ventilation systems
表 3.3 商用厨房通风系统的噪声源

Source of Noise 噪声源	How/Why Noise Arises 噪声产生的原因
Extract hood 排风罩	-High air velocities through extract hood – 穿过排风罩的高速气流
Extract/supply grille 排气 / 进气格栅	-High air velocities through extract/supply grille – 穿过排气 / 进气格栅的高速气流
Extract/supply ductwork 排气 / 进气管道	-High air velocities through extract/supply ductwork -Resonance of fan noise through extract/supply ductwork – 穿过排气 / 进气管道的高速气流 – 穿过排气 / 进气管道的风扇共振
Extract/supply fan 排气 / 进气扇	-Fan motor noise -Fan impeller turning – 风扇电机噪声 – 风扇叶轮转动
Extract/supply discharge point 排气 / 进气排放点	Convictions – 高速气流

3.2 Ventilation Systems (See Figure 3.3)

A number of Local Authorities have been contacted to review the types of problems encountered by Council Officers when dealing with odour and noise situations. Responses were received from metropolitan and rural Authorities. Authorities from England, Northern Ireland, Scotland and Wales have been consulted. The main areas of concern are summarised in Table 3.4

3.2 通风系统（见图 3.3）

一些地方当局回顾了在处理气味和噪声公害时所遇到的问题类型，从城市和乡村当局获得了反馈，涉及英格兰、北爱尔兰、苏格兰和威尔士的地方当局。表 3.4 总结了主要涉及的领域。

Chapter 3: Indoor Air Climate in Open Kitchen Restaurants

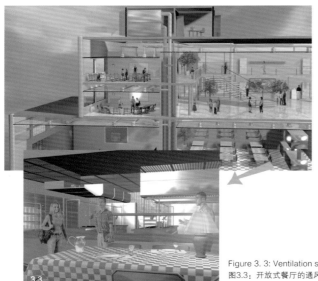

Figure 3. 3: Ventilation system in open-kitchen restaurant
图3.3：开放式餐厅的通风系统

Table 3.4 Summary of problems of commercial kitchen ventilation systems
表 3.4 商用厨房通风系统的问题总结

Area 领域	Effect 效果
Where restaurant changes cooking type (e.g. from tea room to fish and chip shop) 餐厅改变烹饪类型（例如，从茶室改为炸鱼和薯条店）	Generally found that change introduces more extensive odour emissions over longer opening times. Existing planning permission may not include an odour control requirement. Need to rely on nuisance legislation to ensure mitigation measures are installed. 这种变化会在营业时间内产生更多的气味。 原有的建筑许可可能不包含气味控制要求。需要依赖相关法规来保证餐厅采取缓解办法。
Application of carbon filtration 应用碳过滤法	Mixed experience with carbon filters. Concerns about: · Poor maintenance; · Effect on back pressure leading to noise; · Effect on fan size leading to noise; and · Maintenance interval of 4 to 6 months considered appropriate always used in conjunction with stack for discharge. 与碳过滤器混合使用，相关问题包括： · 不良的维护 · 反压效果导致噪声 · 风扇规格导致噪声 · 维护周期在 4～6 个月，通常与排气筒联合使用
UV/ozone systems 紫外线/臭氧系统	Limited experience with these systems. Concerns about: · Residual ozone always used in conjunction with high efficiency particulate removal and a stack. 此类系统的应用有限，相关问题包括： · 残留的臭氧通常与高效悬浮微粒去除器和排气筒联合使用

Application of odour neutralising agents 应用气味中和剂	Mixed experiences with this range of products. Concerns about: · Dosing levels leading to further odour problems; · On going maintenance. Can be used as a "polishing" technique in highly sensitive situations. Must be used in conjunction with stack. 此类产品的混合使用，相关问题包括： · 配量会导致进一步的气味问题 · 持续的维护 可以被应用为高度敏感场合的"精修"技术 必须与排气筒联合使用。
Application of stack height 排气筒高度的应用	Mixed experience with stack heights. No consistency on appropriate height (roof eaves or ridge). Main concern occurred where: · Premises on rising ground where effective stack height is reduced; · Building housing premises is shorter than surrounding buildings; and · Premises is a listed building, is located in a conservation area or located in a courtyard. 排气筒高度的混合应用，没有稳定的合适高度（屋檐或屋脊），相关问题包括： · 对建于高地上的房屋来说，排气筒高度的有效性会削弱 · 建造房屋低于周边建筑 · 不适用于文物保护建筑，或位于保护区内、庭院内的建筑

The commercial kitchen is a unique space where many different HVAC applications take place within a single environment. Exhaust, supply, transfer, refrigeration, building pressurisation and air conditioning all must be considered in the design of most commercial kitchens.

It is obvious that the main activity in the commercial kitchen is the cooking process. This activity generates heat and effluent that must be captured and exhausted from the space in order to control odour and thermal comfort. The kitchen supply air, whether mechanical or transfer or a combination of both, should be of an amount that creates a small negative pressure in the kitchen space. This will avoid odours and

商用厨房是一个特殊的空间，多种不同的空调设备聚集在这个单一环境内。排气、进气、换气、冷藏、建筑增压和空调都必须纳入商用厨房的设计范畴内。

很明显，商用厨房内的主要活动是烹饪。烹饪所产生热量和废气必须被排出去，以实现气味和热舒适度的控制。厨房的供气，无论是机械供气、换气或是混合供气，都应达到一定的量，从而在厨房空间内形成小范围的负压。这能避免气味和污染的空气

Chapter 3: Indoor Air Climate in Open Kitchen Restaurants

Figure 3.4: Heat gain and emission inside the kitchen
图3.4：厨房的热增量和排放

contaminated air escaping into surrounding areas. Therefore the correct exhaust air flow quantity is fundamental to ensure good system operation, thermal comfort and improved IAQ.

Similar considerations should be given to washing-up, food preparation and serving areas.

Initial Design Considerations
The modes of heat gain in a space may include solar radiation and heat transfer through the construction together with heat generated by occupants, lights and appliances and miscellaneous heat gains as air infiltration should also be considered.

Sensible heat (or dry heat) is directly added to the conditioned space by conduction, convection and radiation. Latent heat gain occurs when moisture is added to the space (e.g., from vapour emitted by the cooking process, equipment and occupants). Space heat gain by radiation is not immediate. Radiant energy must first be absorbed by the surfaces that enclose the space (walls, floor, and ceiling) and by the objects in the space (furniture, people, etc.). As soon as these surfaces and objects become warmer than the space air, some of the heat is transferred to the air in the space by convection. (See Figure 3.4).

To calculate a space cooling load, detailed building design information and weather data at selected design conditions are required. Generally, the following information is required:
· building characteristics
· configuration (e.g, building location)
· outdoor design conditions
· indoor design conditions
· operating schedules
· date and time of day

However, in commercial kitchens, cooking processes contribute the majority of heat gains in the space.

释放到周边区域。因此，合适的排气流量对保证良好的系统运营、热舒适度和室内空气质量来讲十分重要。

餐具洗涤区、食品准备区和服务区同样应当注意以上事项。

初步设计要点
一个空间的热增量模式可能包括太阳辐射，建筑结构的热传递，使用者、电灯、电器所产生的热量以及空气渗透所产生混合热量等。

显热（或干热）通过传导、对流和辐射直接添加到空调空间。潜热则与湿气一起进入空间（例如，烹饪过程、设备和使用者所产生的水蒸气）。空间的辐射热增量不是立即产生的。辐射能必须首先被包围空间的表面（墙壁、地面和天花板）和空间内的物品（家具、人等）吸收。只要这些表面和物体的温度高于室内空间，一些热量就会通过对流传递到空气中。（见图3.4）

要计算空间制冷负荷，需要明确详细的建筑设计信息和特定设计条件下的气候数据。通常需要以下信息：
• 建筑特点
• 配置（如建筑的位置）
• 室外设计条件
• 室内设计条件
• 运营时间安排
• 日期和时间

在商用厨房中，烹饪过程是空间热增量的主要来源。

Heat Gain and Emissions Inside the Kitchen

Cooking can be described as a process that adds heat to food. As heat is applied to the food, effluent is released into the surrounding environment. This effluent release includes water vapour, organic material released from the food itself, and heat that was not absorbed by the food being cooked. Often, when pre-cooked food is reheated, a reduced amount of effluent is released, but water vapour is still emitted to the surrounding space.

The hot cooking surface (or fluid, such as oil) and products create thermal air currents (called a thermal plume) that are received or captured by the hood and then exhausted. If this thermal plume is not totally captured and contained by the hood, they become a heat load to the space.

There are numerous secondary sources of heat in the open kitchen (such as lighting, people, and hot meals) that contribute to the cooling load as presented in Table 3.5.

厨房内的热增量和热排放

烹饪可以认为是给食物添加热量的过程。随着热量被添加到食物中，废气也释放到了周边环境里。释放的废气包括水蒸气、食物本身的有机物和食物在烹饪中未吸收的热量。通常，熟食的加热过程所产生的废气会较少，但是水蒸气仍会排放到周边空间之中。

排风罩会吸取热烹饪面（或液体，如油）和食品会产生热空气流（称为热卷流）并将其排出。如果热卷流没有被排风罩完全吸收和截留，就会称为空间的热负荷。

开放式厨房有多种次级热源（如照明、人和热餐），会增加空间的制冷负荷，详见表3.5。

Table 3.5 Cooling load from various sources
表 3.5 各种来源的制冷负荷

Load 负荷	W 功率（瓦）
Lighting 照明	$21\sim54/m^2$ 21~54/平方米
People 人	130/person 130/人
Hot meal 热餐	15/meal 15/餐
Cooking equipment 烹饪设备	Varies 变量
Refrigeration 冷藏设施	Varies 变量

Chapter 3: Indoor Air Climate in Open Kitchen Restaurants

Thermal Comfort, Productivity and Health

Thermal Comfort
One reason for the low popularity of kitchen work is the unsatisfactory thermal conditions.

Thermal comfort is a state where a person is satisfied with the thermal conditions.

The International Organisation for Standardisation (ISO) specifies such a concept as the predicted percentage of dissatisfied occupants (PPD) and the predicted mean vote (PMV) of occupants. PMV represents a scale from -3 to 3, from cold to hot, with 0 being neutral. PPD tells what percentage of occupants are likely to be dissatisfied with the thermal environment. These two concepts take into account four factors affecting thermal comfort: (See Figure 3.5)
· air temperature
· radiation
· air movement
· humidity

The percentage of dissatisfied people remains under 10% in neutral conditions if the vertical temperature difference between the head and the feet is less than 3°C and there are no other non-symmetrical temperature factors in the space. A temperature difference of 6-8°C increases the dissatisfied percentage to 40-70%.

There are also important personal parameters influencing the thermal comfort (typical values in kitchen environment in parenthesis):
· clothing (0.5 - 0.8 clo)
· activity (1.6 - 2.0 met)

Clo expresses the unit of the thermal insulation of clothing (1 clo = 0.155 m^2 K/W).

热舒适度、生产力与健康

热舒适度
人们不愿在厨房工作的原因之一就是其令人不舒服的热度。

热舒适度是人体对热状况感到舒适的状态。

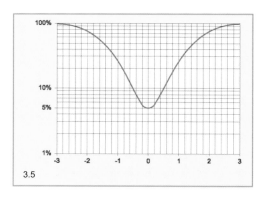

Figure 3.5: PPD as a function of PMV
图3.5：不满意使用者的预测比例和使用者的预测指数

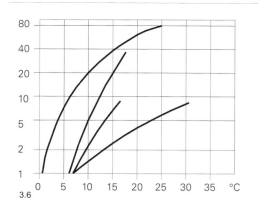

Figure 3.6: Assymmetric thermal radiation
图3.6：非对称热辐射

Met is a unit used to express the metabolic rate per unit Dubois area, defined as the metabolic rate of a sedentary person, 1 met = 50 kcal/(hr. m^2) = 58.2 W/m^2.

Assymmetric Thermal Radiation
In the kitchen, the asymmetry of radiation between the cooking appliances and the surrounding walls is considerable as the temperature difference of radiation is generally much higher than 20° C.（See Figure 3.6）

Ventilation Effectiveness and Air Distribution System

The Effect of Air Supply
Ventilation effectiveness can be described as the ability of ventilation system to achieve design conditions in the space (air temperature, humidity, concentration of impurities and air velocity) at minimum energy consumption. Air distribution methods used in the kitchen should provide adequate ventilation in the occupied zone, without disturbing the thermal plume.

In the commercial kitchen environment the supply airflow rate required to ventilate the space is a major factor contributing to the system energy consumption. Traditionally high velocity mixing or low velocity mixing systems have been used. Now there is a third alternative that clearly demonstrates improved thermal comfort over mixing systems, this is displacement ventilation.

The supply air (make-up air) can be delivered to the kitchen in two ways:
· high velocity or mixing ventilation
· low velocity or displacement

Low Velocity or Displacement Ventilation

国际标准化组织详细说明了不满意使用者的预测比例和使用者的预测指数。预测指数表示为 –3 ~ 3，从冷到热，0 为中间数。不满意使用者的预测比例指出了对热环境不满意的使用者的百分比。这两个概念从四个方面影响热舒适度（见图3.5）：
• 气温
• 辐射
• 气流
• 湿度

如果垂直温度差小于3℃且空间内没有其他非对称温度因素，不满意人群的百分比可保持在10%以下。6 ~ 8℃的温度差会让不满意人群的百分比上升至40% ~ 70%。

还有一些影响热舒适度的个人参数（厨房环境的参考数据）：
• 衣物（0.5 ~ 0.8 克罗）
• 活动（1.6 ~ 2.0 米特）

克罗（clo）是表示衣物隔热效率的单位，1 clo = 0.155 m^2 K/W。

米特（met）是表示单位体表面积代谢效率的单位，定义为人坐着时的新陈代谢率，1 met = 50 kcal/(hr. m^2) = 58.2 W/m^2。

非对称热辐射
在厨房中，炊具与四周墙壁之间的非对称热辐射相当可观，辐射的温度差通常远高于20℃。（见图3.6）
通风效率与空气分配系统

Chapter 3: Indoor Air Climate in Open Kitchen Restaurants

Here, the cooler-than-surrounding supply air is distributed with a low velocity to the occupied zone. In this way, fresh air is supplied to where it is needed. Because of its low velocity, this supply air does not disturb the hood function. (See Figure 3.7)

High velocity or Mixing Ventilation
Everything that is released from the cooking process is mixed with the supply air. Obviously impurities and heat are mixed with surrounding air. Also the high velocity supply air disturbs the hood function. (See Figure 3.8) With a displacement system the intensity of turbulence of about 10 %, one accepts velocities between 0.25 and 0.40 m/s, with the air between 20 and 26°C respectively with 20% of people dissatisfied.

In the case of mixing ventilation, with an intensity of turbulence from 30 to 50 %, one finds 20 % of people dissatisfied in the following conditions: (See Table 3.6)

Productivity
Room air temperature affects a person's capacity to work. Comfortable thermal conditions decrease the number of accidents occurring in the work place. When the indoor temperature is too high (over 28 °C in commercial kitchens) the productivity and general comfort diminish rapidly.

The average restaurant spends about $2,000 yearly on salaries in the USA, wages and benefits per seat. If the air temperature in the restaurant is maintained at Picture 6. Productivity vs. Room Air Temperature 27°C in the kitchen the productivity of the restaurant employees is reduced to 80 %. That translates to losses of about $40,000 yearly on salaries and wages for an owner of a 100-seat restaurant. (See Figure 3.9)

Health
There are several studies dealing with cooking and health issues. The survey confirmed that cooking fumes contain hazardous components in both Western and Asian types of kitchens. In one study, the fumes

空气供给的效果
通风效率可被描述为通风系统在最低能源消耗下实现空间设计条件（气温、湿度、杂质浓度和气流速度）的能力。厨房的空气配送方式应当提供足够的通风，同时又不影响热卷流。

在商用厨房环境中，空间通风所需的进气流速是影响系统能源消耗的主要因素。传统上习惯使用高速混合系统或低速混合系统。现在有一种新方法，可以显著提高热舒适度，即置换通风。

供给空气（补充空气）可以通过两种方式传送至厨房：
- 低速或置换通风
- 高速或混合通风

低速或置换通风
温度低于环境空气的供气以低速配送到使用区域。这样一来，有需要的区域就获得新鲜空气。由于速度较低，这种空气供给不会影响排风罩的功能使用。（见图 3.7）

第三章　开放式餐厅的室内空气环境

Figure 3. 7: Low velocity or displacement ventilation
With a displacement system the intensity of turbulence of about 10 %, one accepts velocities between 0.25 and 0.40 m/s, with the air between 20 and 26°C respectively with 20% of people dissatisfied.
图3.7：低速通风或置换通风
假设置换系统的乱流强度在10%左右，人体可接受的气流速度在0.25～0.40m/s，气温在20～26°C时，不满意人群可达到20%。

Figure 3.8: High velocity or mixing ventilation
In the case of mixing ventilation, with an intensity of turbulence from 30 to 50 %, one finds 20 % of people dissatisfied in the following conditions:
图3.8：在混合通风中，假设乱流强度为30%～50%，20%的人在以下情况下会感到不满意：

Table 3.6 Air temperature/air velocity
表3.6 气温/气流速度

Air temperature. (°C) 气温（°C）	20	26
Air velocity (m/s) 气流速度（m/s）	0.15	0.25

高速或混合通风
烹饪过程所释放的所有物质都将与供给空气相混合。杂质和热量都混合在环境空气中。高速供气还会妨碍排风罩的功能使用。（见图3.8）

假设置换系统的乱流强度在10%左右，人体可接受的气流速度在0.25～0.40m/s，气温在20°C～26°C时，不满意人群可达到20%。
在混合通风中，假设乱流强度为30%～50%，20%的人在以下情况下会感到不满意：（见表3.6）

生产力
室内气温影响着人的工作能力。舒适的热状况能够减少工作事故的发生。当室内温度过高（商用厨房中高于28°C）时，生产力和总舒适度会急速降低。美国餐厅每年的薪水和福利开支平均到每个餐位上是2,000美元。如果餐厅厨房的气温保持在27°C，餐厅的员工将减少至80%。对于100个餐位的餐厅的经营者来说，就相当于每年损失40,000美元。（见图3.9）

Chapter 3: Indoor Air Climate in Open Kitchen Restaurants

generated by frying pork and beef were found to be mutagenic. In Asian types of kitchens, a high concentration of carcinogens in cooking oil fumes has been discovered. All this indicates that kitchen workers may be exposed to a relatively high concentration of airborne impurities and that cooks are potentially exposed to relatively high levels of mutagens and carcinogens.

Chinese women are recognised to have a high incidence of lung cancer despite a low smoking rate e.g. only 3% of women smoke in Singapore. The studies carried out show that inhalation of carcinogens generated during frying of meat may increase the risk of lung cancer.

The risk was further increased among women stirfrying meat daily whose kitchens were filled with oily fumes during cooking. Also, the statistical link between chronic coughs, phlegm and breathlessness on exertion and cooking were found.

In addition to that, Cinni Little states, that three quarters of the population of mainland China alone use diesel as fuel type instead of town gas or LPG, causing extensive bronchial and respiratory problems among kitchen workers, which is possibly exacerbated by an air stream introduced into the burner mix.

Reduction of Health Impact

The range of thermal comfort neutrality acceptable without any impact on health has been proposed as running between 17°C as the lowest and 31°C as the highest acceptable temperature (Weihe 1987, quoted in WHO 1990). Symptoms of discomfort and health risks outside this range are indicated in Table 3.7.

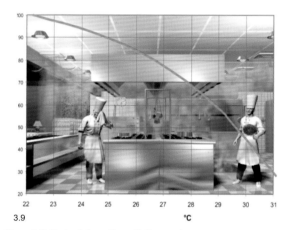

Figure 3. 9: Productivity vs. Room Air Temperature
图3.9：生产力与室内气温

健康
学者针对烹饪和健康做了一些研究。调查显示东西方厨房的烹饪油烟中都包含有害物质。一项研究表明，炸猪肉和牛肉的油烟具有致突变性。研究还发现，亚洲式厨房的烹饪油烟中有高浓度致癌物。这些都表明厨房工作者可能暴露在相对较高浓度的空气杂质中，而厨师则暴露在相对较高等级的致突变物和致癌物中。

尽管华裔女性的吸烟率极低（新加坡女性的吸烟率仅为3%），她们罹患肺癌的可能性却较高。研究表明，吸入油炸肉类所产生的致癌物会增加罹患肺癌的风险。

如果女性进行烹饪的厨房内每天充满油炸肉类所产生的油烟，则风险会进一步增加。同时，统计数据显示，慢性咳嗽、痰和呼吸急促等问题也与烹饪有关。

Table 3.7 Health effects of thermal microclimates lying outside the neutral comfort zone
表 3.7 舒适范围外的热环境对健康的影响

	<17°C	>31°C	
Sudden heart death 心脏突死	Hypertension 高血压	Hypotension 高血压	Sudden heart death 心脏突死
Stroke 中风	Hypertension 高血压	Hyperthermia 过高热	Heart failure 心力衰竭
	Respiratory infections 呼吸道感染	Tachycardia 心动过速	Heat stroke 中暑
	Asthma 哮喘	Health insufficiency 亚健康	
	Overheating 过热	Inappetence 食欲不振	
	Tachycardia 心动过速	Hypohydrosis 多汗	
	Reduced dexterity 灵敏度下降	Hydromeiosis 代谢过度	
	Indolecense 迟缓	Indolence 懒惰	
	Restlessness 坐立不安	Fatigue 疲劳	
	Mental slowing 精神迟钝	Lethargy 昏睡	
	Depression 抑郁	Increased irritability 易怒	

Chapter 3: Indoor Air Climate in Open Kitchen Restaurants

Ventilation Rate
The airflow and air distribution methods used in the kitchen should provide adequate ventilation in the occupied zone, without disturbing the thermal plume as it rises into the hood system. The German VDI-2052 standard states that a:
Ventilation rate over 40 vol./h result on the basis of the heat load, may lead to draughts.

The location of supply and exhaust units are also important for providing good ventilation. Ventilating systems should be designed and installed so that the ventilation air is supplied equally throughout the occupied zone. Some common faults are to locate the supply and exhaust units too close to each other, causing "short-circuiting" of the air directly from the supply opening to the exhaust openings. Also, placing the high velocity supply diffusers too close to the hood system reduces the ability of the hood system to provide sufficient capture and containment (C&C) of the thermal plume.

Recent studies show that the type of air distribution system utilised affects the amount of exhaust needed to capture and contain the effluent generated in the cooking process.

Integrated Approach
Energy savings can be realised with various exhaust hood applications and their associated make-up air distribution methods. However with analysis the potential for increased energy savings can be realised when both extract and supply for the kitchen are adopted as an integrated system.

The combination of high efficiency hoods and displacement ventilation reduces the required cooling capacity, while maintaining temperatures in the occupied space. The natural buoyancy characteristics of the displacement air helps the C&C of the contaminated convective plume by 'lifting' it into the hood. (See Figure 3.10)

减轻对健康的影响
不影响健康的热舒适度的可接受范围在17℃到31℃之间（Weihe 1987, quoted in WHO 1990）。在这一范围外所引起的不舒适症状和健康风险详见表3.7。

通风率
厨房的气流和空气配送方式应当提供足够的通风，同时又不影响排风系统的热卷流。德国标准VDI-2052显示，通风率大于40 vol./h时会引起气流。

进气和排气机组的位置对提供良好的通风同等重要。通风系统应当进行良好的设计和安装，让使用区域的各个角度都能得到通风供给。进气和排气机组的位置过近或导致气流"短路"，使空气直接从进气口到达排气口。此外，将高速供气分散口设置得离排风系统过近会降低后者的效率，使其不足以吸收和控制热卷流。

第三章　开放式餐厅的室内空气环境

Third-party research has demonstrated that this integrated approach for the kitchen has the potential to provide the most efficient and lowest energy consumption of any kitchen system available today. (See Figure 3.10)

最近的研究表明，所采用的空气配送系统类型会影响吸收和控制烹饪废气的排气装置的数量。

综合方法
运用各种排风罩和配套的补气分配方法可以实现节能。要进一步增加节能量，可以在厨房采用排气和供气的综合系统。

高效排风罩和置换通风的组合能够减少制冷负荷，同时还能保持空间的温度。新鲜自然的置换空气有助于排风罩吸收和控制污染的对流油烟。

研究表明，在厨房中应用这种综合方法能够实现厨房系统的高效节能。（见图3.10）

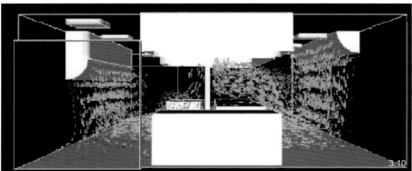

Figure 3.10: Displacement ventilation
图3.10：置换通风

3.3 Kitchen Hoods

The purpose of kitchen hoods is to remove the heat, smoke, effluent, and other contaminants. The thermal plume from appliances absorbs the contaminants that are released during the cooking process. Room air replaces the void created by the plume. If convective heat is not removed directly above the cooking equipment, impurities will spread throughout the kitchen, leaving discoloured ceiling tiles and greasy countertops and floors. Therefore, contaminants from stationary local sources within the space should be controlled by collection and removal as close to the source as is practical.

Appliances contribute most of the heat in commercial kitchens. When appliances are installed under an effective hood, only the radiant heat contributes to the HVAC load in the space. Conversely, if the hood is not providing sufficient capture and containment, convective and latent heat are "spilling" into the kitchen thereby increasing both humidity and temperature.

Capture efficiency is the ability of the kitchen hood to provide sufficient capture and containment at a minimum exhaust flow rate.

3.3.1 Selecting & Sizing Exhaust Hoods
Fundamentals of Kitchen Exhaust

An exhaust fan in the ceiling could remove much of the heat produced by cooking equipment. But mix in smoke, volatile organic compounds, grease particles and vapor from cooking, and a means to capture and contain the effluent becomes necessary to avoid health and fire hazards. While an exhaust hood serves that purpose, the key question becomes: what is the appropriate exhaust rate? The answer always depends on several factors: the menu of food products and the type (and use) of the cooking equipment under the hood, the style and geometry of the hood itself, and how the makeup air (conditioned or otherwise) is introduced into the kitchen.

3.3 厨房排风罩

厨房排风罩的作用是去除热气、烟雾、废气和其他污染物。排风罩的热卷流吸收烹饪过程所产生的污染物。室内空气补充热卷流造成的真空区域。如果没有及时去除烹饪设备上方的对流热，杂质会扩散到厨房的各个角落，污染天花板、台面和地面。因此，由固定位置所产生的污染物应当被尽可能地快速、就近去除。

商用厨房中的炊具是主要的热源。如果将炊具安装在高效的排风罩下方，就只有辐射热会造成空调系统的负荷。相反，如果排风罩不能有效地吸收并控制热气，对流热和潜热就会分散到厨房的各个角落，从而导致湿度和温度的上升。

吸收效率是厨房排风罩在最低排气流量下提供充分吸收和控制的能力。

3.3.1 排风罩的选择与尺寸
厨房排风的基本要求
天花板上的排风扇可以去除炊具产生的大部分热量。烹饪产生的混合烟雾、挥发性有机物、油脂颗粒和水蒸气需要吸收和控制，以避免产生健康问题和火灾危险。排风罩起到了这一作用，但是主要问题是：如何确定合适的排气率？这取决于几个因素：食品菜单、排风罩下炊具的类型和使用、排风罩的风格和形状以及厨房的空气补给方式（空调或其他方式）。

The Cooking Factor

Cooking appliances are categorized as light-, medium-, heavy-, and extra heavy-duty, depending on the strength of the thermal plume and the quantity of grease, smoke, heat, water vapor, and combustion products produced. The strength of the thermal plume is a major factor in determining the exhaust rate. By their nature, these thermal plumes rise by natural convection, but they are turbulent and different cooking processes have different "surge" characteristics. For example, the plume from hamburger cooking is strongest when flipping the burgers. Ovens and pressure fryers may have very little plume until they are opened to remove food product. Open flame, non-thermostatically controlled appliances, such as underfired broilers and open top ranges, exhibit strong steady plumes. Thermostatically controlled appliances, such as griddles and fryers have weaker plumes that fluctuate in sequence with thermostat cycling (particularly gas-fired equipment). As the plume rises, it should be captured by the hood and removed by the suction of the exhaust fan. Air in the proximity of the appliances and hood moves in to replace it. This replacement air, which must ultimately originate as outside air, is referred to as makeup air.

Building codes distinguish between cooking processes that create smoke and grease (e.g., frying, griddling, or charbroiling) and those that produce only heat and moisture (e.g., dishwashing and some baking and steaming operations). Cooking that produces smoke and grease requires liquid-tight construction with a built-in fire suppression system (Type I hood), while operations that produce only heat and moisture do not require liquid-tight construction or a fire sup-pression system (Type II hood).

Menu items may produce more or less smoke and grease depending on their fat content and how they are cooked. Higher fat content foods tend to re-lease more smoke and grease regardless of the type of cooking process. Testing under an ASHRAE sponsored research project at the University of Minnesota confirmed that hamburger cooked on a charbroiler releases finer smoke parti-cles and more grease vapour and particles than hamburger cooked on a griddle. The percentage fat content

烹饪因素

根据热卷流的强度以及油脂、烟、热、水蒸气和燃烧产物的数量，可以将炊具分为几大类：低排放、中排放、高排放和超高排放。热卷流的强度是决定排风率的主要因素。从本质上讲，热卷流通过自然对流而上升，但是它们是混乱的，且不同的烹饪过程会产生不同形式的热卷流。例如，翻炒汉堡牛肉饼所产生的卷流是最强的。烤箱和压力炸锅只有在开盖的时候才产生大量卷流。明火、非恒温控制炊具（如下加热式烘烤机和开盖炊具）会产生强烈而稳定的卷流。恒温控制炊具（如平底锅和煎锅）产生的卷流较弱，且会随着恒温循环（特别是燃气设备）而波动。随着卷流的上升，排风罩会吸入卷流，并通过排风扇将其排出。炊具和排风罩周围的空气将会置换卷流的位置。这种置换空气源于室外，被称为补充空气。

建筑规范将烹饪流程分为两类：一类产生油烟（如油炸、煎或炭烤），另一类只产生热气和水蒸气（如烘焙、蒸煮）。产生油烟的烹饪流程需要内置灭火系统的液封结构风罩（I型排风罩），而只产生热气和水蒸气的烹饪流程则无需液封结构和灭火系统（II型排风罩）。

菜品所产生的油烟量取决于其自身的脂肪含量和烹饪方式。高脂肪含量的食品会释放更多的油烟。明尼苏达州大学一项由美国采暖、制冷与空调工程师协会（ASHRAE）所支持的研究证明：与在平底锅上烹饪相比，在烧烤炉烹饪汉堡牛肉饼会产生更细的烟雾颗粒和更多的油烟。牛肉饼的脂肪含量也会影响烹饪所产生的油烟量。鸡胸肉的脂肪含量低于牛肉饼，所以它在烧烤炉和平底锅上烹饪所产生的

Chapter 3: Indoor Air Climate in Open Kitchen Restaurants

of hamburger also contributes to differences in the amount of grease and smoke released in cooking. Chicken breast, which has less fat compared to hamburger, releases less particulate and less grease during cooking on a charbroiler or on a griddle compared to hamburger.

The Hood Factor

The design exhaust rate also depends on the hood style and construction features. Wall-mounted canopy hoods, island (single or double) canopy hoods, and proximity (backshelf, pass-over, or eyebrow) hoods all have different capture areas and are mounted at different heights and horizontal positions relative to the cooking equipment. Generally, for the identical (thermal plume) challenge, a single-island canopy hood requires more exhaust than a wall-mounted canopy hood, and a wall-mounted canopy hood requires more exhaust than a proximity (backshelf) hood. The performance of a double-island canopy tends to emulate the performance of two back-to-back wall-canopy hoods, although the lack of a physical barrier between the two hood sections makes the configuration more susceptible to cross drafts.

Building and/or health codes typically provide basic construction and materials requirements for exhaust hoods, as well as prescriptive exhaust rates based on appliance duty and length of the hood (cfm per linear ft.) or open face area of the hood (cfm per ft^2). Codes usually recognize exceptions for hoods that have been tested against a recognized standard, such as Underwriters Laboratories (UL) Standard 710. Part of the UL standard is a "cooking smoke and flair up" test. This test is essentially a cooking effluent capture and containment (C&C) test where "no evidence of smoke or flame escaping outside the exhaust hood" must be observed. Hoods bearing a recognized laboratory mark are called listed hoods, while those constructed to the prescriptive requirements of the building code are called unlisted hoods. Generally, an off-the-shelf listed hood can be operated at a lower exhaust rate than an unlisted hood of comparable style and size over the same cook line. Lower exhaust rates may be proven by labora-tory testing with specific hood(s) and appliance lineup using the test protocol described in ASTM Standard

油烟都低于牛肉饼。

排风罩因素

设计排风率还取决于风罩类型及其结构特点。壁挂式伞形排风罩、岛式伞形排风罩、近距离排风罩的吸收面各不相同，且安装在不同的高度及炊具相对水平位置。通常来讲，面对同等强度的热卷流，单岛式伞形排风罩的排风率要高于壁挂式伞形排风罩，而壁挂式伞形排风罩的排风率又高于近距离排风罩。尽管两个风罩之间缺乏物理屏障，易受交叉气流影响，一个双岛伞形排风罩的效率还是接近于两个背对式壁挂伞形排风罩。

建筑规范和健康守则通常都提出了排风罩的基本结构和材料要求以及对应炊具的标准排风率和风罩长度或开放吸收面积。如果排风罩经过权威标准的检测，规范也会认可其标准，如美国保险商实验室（UL）标准710的"烹饪油烟测试"。这一测试的本质

F-1704, Test Method for Performance of Commercial Kitchen Ventilation Systems. This process is sometimes referred to as "custom-engineering" a hood.

Laboratory testing of different combinations of appliances has demonstrated that minimum capture and containment rates vary significantly due to appliance type and position under the hood. For example a heavy-duty appliance at the end of a hood is more prone to spillage than the same appliance located in the middle of the hood. (See Figure 3.11)

是吸收和控制废气能力测试。具有实验室认证标签的排风罩称为认证排风罩，而根据建筑规范所配备的排风罩则称为非认证排风罩。通常来讲，现成的认证排风罩的运行转速要低于同类型的非认证排风罩。美国材料试验协会（ASTM）标准F-1704的"商用厨房通风系统性能测试"可以对低排风率进行认证。这一流程也被称为排风罩的"定制工程化"。

实验室对于各种炊具组合的测试表明：吸收和控制率的最小值受炊具类型和排风罩位置的影响。例如，位于排风罩一端的重型炊具比位于排风罩中央的同类型炊具更易溢出油烟。（见图3.11）

侧板和顶罩
侧板（端板）通常可以允许排风罩减小通风率，因为所有置换空气都被吸到设备前方，提高了对废气卷流的控制力。这是一种性价比较高的提升排风能力和减少总排风率的方式。端板的另一个好处是可以缓解交叉气流的负面效应。重要的是，局部侧板的效果几乎与全挡板的效果相同。端板能够显著提升单岛和双岛式伞形排风罩的效率。

顶罩的增大应当提升伞形排风罩的吸收能力，因为它增加了热卷流与风罩边缘的距离。这样一来，就可以实现将炊具设置在排风罩远处的合适位置，还可以增加其边长。尽管之中做法可以提升吸收和控制效率，但是必须相应地增大排风率。建议在产生大量卷流的炊具（如对流和组合烤炉、蒸锅、压力炸锅等）上使用大型顶罩。（见图3.12）

3.11

Figure 3.11: Styles of exhaust hoods
图3.11：排风罩的类型

Chapter 3: Indoor Air Climate in Open Kitchen Restaurants

Side Panels and Overhang

Side (or end) panels permit a reduced ex-haust rate in most cases, as all of the replacement air is drawn across the front of the equipment, which improves containment of the effluent plume generated by the hot equipment. They are a relatively inexpensive way to improve C&C and reduce the total exhaust rate. Another benefit of end panels is to mitigate the negative effect that cross drafts can have on hood performance. It is important to know that partial side panels can provide almost the same benefit as full panels. Although tending to defy its definition as an "island" canopy, end panels can improve the performance of a double-island or single-island canopy hood.

An increase in overhang should improve the ability of a canopy hood to capture because of the increased distance between the plume and hood edges. This may be accomplished by pushing the appliances as far back under a canopy hood as practical and/or by increasing the side length. Although this improves C&C performance, for unlisted hoods under a local jurisdiction referencing the Uniform Mechanical Code (UMC), this would require increase in the code-required exhaust rate. Larger overhangs are recommended for appliances that create plume surges, such as convection and combination ovens, steamers and pressure fryers. This was the driving argument for converting the code-specified exhaust rates from a "cfm/ft^2" to a "cfm/linear ft". basis in the cur-rent edition of the International Mechanical Code (IMC). (See Figure 3.12)

Hood Geometry

The ability of a hood to capture and contain cooking effluent can often be enhanced by adding passive features (e.g., angles, flanges, or geometric flow deflectors) or active features (e.g., low-flow, high-velocity jets) along the edges of the hood or within the hood reservoir. Such design features can improve hood performance dramatically over a basic box-style hood with the same nominal dimensions.

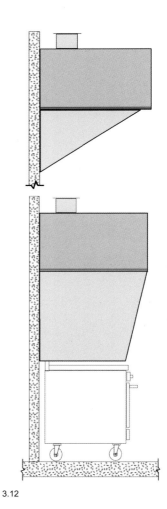

3.12

Figure 3.12: Illustration of partial and full side panels
图3.12：局部侧板与全挡板的示意图

Cross Drafts

Cross drafts can have a detrimental affect on all hood/appliance combinations. Cross drafts affect island canopy hoods more than wall mounted canopy hoods because they have more open area allowing drafts to push or pull effluent from the hood. For example, a pedestal fan used by staff for additional cooling can severely degrade hood performance, may make capture impossible and may spill the plume into the kitchen. Location of delivery doors, service doors, pass-through openings and drive-through windows may be sources of cross drafts due to external and internal air pressure differences. Cross drafts can also be developed when the makeup air system is not working correctly, causing air to be pulled from open drive-through windows or doors.

The Makeup Factor

The layout of the heating, ventilating, and air-conditioning (HVAC) and makeup air (MUA) supply air outlets or diffusers can affect hood performance. These can be sources that disrupt thermal plumes and hinder C&C. Safety factors are typically applied to the design exhaust rate to compensate for the effect that undesired air movement within the kitchen has on hood performance.

Air that is removed from the kitchen through an exhaust hood must be replaced with an equal volume of outside replacement (makeup) air through one or more of the following pathways:
1. Transfer air (e.g., from the dining room)
2. Displacement diffusers (floor or wall mounted)
3. Ceiling diffusers with louvers (2-way, 3-way, 4-way)
4. Slot diffusers (ceiling)
5. Ceiling diffusers with perforated face
6. Integrated hood plenum including:
   · Short circuit (internal supply)
   · Air curtain supply
   · Front face supply
   · Perforated perimeter supply
   · Backwall supply (rear discharge)

排风罩结构
可以通过在排风罩边缘或储藏器内添加被动特征（例如棱角、凸缘、导流板）或主动特征（例如低流量的高速喷射器）来提升排风罩吸收和控制烹饪废气的能力。这些设计特征能够在同等尺寸范围内显著提高基本盒式排风罩的性能。

交叉气流
交叉气流会影响排风罩和炊具组合装备的效率。交叉气流对岛式伞形排风罩的影响要大于对壁挂式伞形排风罩的影响，因为前者的开放面积更大，更便于气流将废物推进或拉出排风罩。例如，员工所使用的落地扇会严重影响排风罩的效率，可能会导致排风罩失效，让油烟散布在厨房中。由于内外空气的压力差，送货门、服务门、传菜口和打包窗口都可能是交叉气流的来源。补气系统的运转错误也可能导致空气从打包窗口进行，从而形成交叉气流。

补充因素
采暖、通风和空调的布局以及补气系统送气口或散流器的位置都能影响排风罩的效率。这些都是影响热卷流和妨碍排风罩吸收和控制能力的来源。设计排风率通常设有安全系数，以抵消厨房内不良空气流动对排风罩效率的影响。

通过排风罩从厨房中排出的空气细部被同等体积的室外置换空气（补偿空气）以下途径置换：
1. 转移空气（例如，从餐厅转移到厨房）
2. 置换散流器（地面式或壁挂式）
3. 栅格式顶棚散流器（双向、三向、四向）
4. 条缝型散流器（顶棚）

Chapter 3: Indoor Air Climate in Open Kitchen Restaurants

· Combinations of the above

The Design Process
Successfully applying the fundamentals of commercial kitchen ventilation (CKV) during the design process requires a good understanding of the local building code requirements, the menu and appliance preferences, and the project's budget. Information about the kitchen equipment and ventilation requirements may evolve over the course of the design phase. Data needed by other members of the design team may require early estimates of certain parameters (e.g., the amount of exhaust and makeup air, motor horsepower, water supply and wastewater flow rates). As more decisions are made, new information may allow (or require) refinements to the design that affect exhaust and makeup air requirements.

The fundamental steps in the design of a CKV system are:
1. Establish location and "duty" classifications of appliances including menu effects. Determine (or coordinate with food service consultant) preferred appliance layout for optimum exhaust ventilation.
2. Select hood type, style, and features.
3. Size exhaust airflow rate.
4. Select makeup air strategy, size airflow and layout diffusers.

A good understanding of how building code requirements apply to kitchen design is essential. Local or state building codes are usually based on one of the "model" building codes promulgated by national code organisations. Our discussion of the building codes will be limited to require-ments that affect design exhaust and makeup air rates, which are usually found in the mechanical code portion of the overall building code.

Historically, codes and test standards used "temperature" ratings for classifying cooking equipment. Although these temperature ratings roughly correlated with the ventilation requirement of the appliances, there were many gray areas. During development of ASHRAE Standard 154, Ventilation for Commercial Cooking Appliances, it was recognised that plume strength, which takes into account plume volume and surge

5. 打孔式顶棚散流器；
6. 综合排风增压系统，包括：
• 短回路（内部供给）
• 气幕供给
• 正面供给
• 打孔式周界供给
• 后墙供给（后方排放）
• 以上综合

设计流程
要想成功地将商用厨房通风的基本原理运用到设计流程中，就必须对地方建筑规范要求、餐厅菜单、厨具类型和项目预算有全面的了解。厨房设备和通风要求的相关信息可能会随着设计的进程而逐步进化。设计团队中其他成员所需的信息可能要求对某些参数进行前期预估（例如，排风率和送风率、电机马力、供水流量和废水流量）。随着决策内容的增加，新信息可能会导致设计的修订，从而影响排气和送气要求。

商用厨房通风系统的主要设计步骤如下：
1. 确定厨房器具的位置和"责任"，包括菜单的影响。确定（或与食品服务顾问协商）优先的器具布局，以优化排风通风。
2. 选择排气罩的型号、风格和特征。
3. 确定排气流量。
4. 选择空气补给策略；确定气流量和散流器的布局。

对建筑规范要求的良好理解在厨房设计中至关重要。地方建筑规范通常以由国家机构发布的某个建筑规范模板为基础。这里所讨论的建筑规范仅限于影响设计排风率和送风率的建筑规范，主要见于整体建筑规范的机械规范部分。

characteristics, as well as plume tempera-ture, would be a better measure for rating appliances for application in building codes. "Duty" ratings were created for the majority of commercial cooking appliances under Standard 154, and these were recently adopted by the International Mechanical Code (IMC). The Kitchen Ventilation chapter of the ASHRAE Applications Handbook (2003 edition) applied the same concept to establish ranges of exhaust rates for listed hoods. The appended Design Exam-ples in this Guide reference duty classifications for appliances. The duty classifications listed in the sidebar are from ASHRAE Standard 154-2003, Ventilation for Commercial Cooking Operations.

The IMC dictates exhaust rates based on hood type and appliance duty. Table 3.8 states these exhaust rates in "cfm per linear foot of hood" ("linear foot" in this case applies to the distance from edge to edge along the front face of the hood). The Code requires that the exhaust rate for the highest duty-rated appliance be applied to the entire hood. The Uniform Mechanical Code (UMC), used in many California jurisdictions, requires calculating exhaust rates based on square-footage of capture area (capture area is the open area defined by the lower edges of the hood). The UMC uses temperature classifications for appliances, as described above. Both the IMC and the UMC require a minimum 6-inch hood overhang (front and sides) for canopy style hoods.

传统上，建筑规范和测试标准都利用"温度"等级来划分烹饪设备。尽管这些温度等级与炊具的通风要求大致相关，仍有许多灰色领域。《ASHRAE标准 154——商用厨房器具的通风》将热卷流强度和热卷流温度作为划分炊具等级的更优标准。"责任"评级主要针对大多数符合标准 154 以及近期被国际机械规范（IMC）所采纳的商用厨房器具。《ASHRAE应用手册（2013 年版）》的厨房通风章节采用同样的概念来划分排风罩的排风率。

IMC 规定，排风率取决于排风罩型号和炊具的热卷流排放量。表 3.8 以"每纵尺的风量"表示排风率。（纵尺指排风罩正面两边之间的距离）。规范要求与排放量最高的炊具相对应排风率应用于整个排风罩。统一机械规范（UMC）要求以排风罩吸收面积为基准计算排风率（吸收面积指排风罩低边的开放面积）。UMC 采用上文提到的温度划分标准。IMC和 UMC 都要求伞形排风罩具有至少为 6 英寸的顶罩（设在前面和侧面）。

Chapter 3: Indoor Air Climate in Open Kitchen Restaurants

Table 3.8 Unlisted Hood Exhaust Flow Rates
表 3.8 未认证排风罩的排风率

Minimum Exhaust Flow Rate for Unlisted Hoods (cfm per linear foot of hood) 未认证排风罩的最低排风率（每纵尺的风量）				
Type of Hood 排风罩类型	Light Duty Equipment 低排放设备	Medium Duty Equipment 中排放设备	Heavy Duty Equipment 高排放设备	Extra-Heavy Duty Equipment 超高排放设备
Wall-mounted Canopy 壁挂式伞形	200	300	400	550
Single Island Canopy 单岛式伞形	400	500	600	700
Double Island Canopy 双岛式伞形	250	300	400	550
Eye Brow 齐眉式	250	250	not allowed 不允许	not allowed 不允许
Backshelf 后架式	250	300	400	not allowed 不允许
Passover 超越式	250	300	400	not allowed 不允许

The prescriptive mechanical code exhaust rate requirements must be conservative because the AHJ (authority having jurisdiction) has no control over the design of an exhaust hood or the positioning and diversity of appliances placed beneath that hood. However, in cases where the CKV system design and appliance configuration has been optimised, the code-specified exhaust rate may be significantly greater than what is required for effective capture and containment of the cooking plume. The code-based safety factor (which may be necessary for unlisted systems)

规定排风率要求必须相对保守，因为权威管辖无法控制排风罩的设计和排风罩下方炊具的摆设。然而，在商用厨房通风系统设计和厨房器具配置优化的前提下，规定排风率会远大于吸收和控制烹饪油烟的需求。规定安全系数（主要适用于未认证的系统）对补偿空气的温度要求可能会增加商用厨房通风系统的能源成本。

can place an energy cost burden on the CKV system through its demand for more heated and cooled makeup air.

When the energy crisis of the 1970's occurred, kitchen ventilation systems became an obvious target. Industry responded with two methods of reducing the amount of replacement air that had to be cooled or heated: (1) short-circuit hoods, and (2) listed hoods.

One strategy, called "internal compensation," was to introduce the makeup air directly into the hood reservoir. This is more commonly known as "short-circuit" makeup air. Although short-circuit hoods have been installed and operated with as much as 80% of replacement air being introduced internally, field and laboratory investigations have shown that these hoods fail to capture and contain effluent adequately.

The second industry strategy was to test hoods under laboratory conditions according to a test protocol specified by Underwriters Laboratories, Standard 710, Exhaust Hoods for Commercial Cooking Equipment. This UL Standard covers materials and construction of exhaust hoods as well as C&C performance. The C&C performance is based on testing a single appliance under a representative hood at one or more of three cooking temperature operating set points (400°F, 600°F, or 700°F). The UL listing reports the minimum C&C rate determined under this laboratory test.

Another American national standard, ASTM Standard F-1704-1999, Test Method for Performance of Commercial Kitchen Ventilation Systems, covers exhaust hood capture and containment performance as well as heat gain from hooded appliances. The current version of ASTM F-1704 also does not address dynamic conditions, but there are amendments under consideration to add a dynamic test that would quantify a safety factor. The capture and containment tests in UL 710 and ASTM F-1704 are similar.

第三章　　开放式餐厅的室内空气环境

在20世纪70年代的能源危机中，厨房通风系统首当其冲。有两种方法能够解决置换空气的加热和制冷问题：(1)短回路排风罩；(2)认证排风罩。

第一种方法称为"内部补偿法"，直接将补偿空气引入排风罩储藏器，俗称"短回路"补给空气。尽管短回路排风罩的安装使用可以从内部引入80%的置换空气，实践调查和实验室研究表明这些排风罩的吸收和控制废气能力不足。

第二种方法是根据《保险商实验室(UL)标准710——商用烹饪设备的排风罩》在实验室对排风罩进行检测。这项UL标准覆盖了排风罩的材料、结构以及性能。其中，对吸收和控制废气性能的检测采用以下方式：以在三个特定烹饪温度（400°F、600°F、700°F）下，典型排风罩针对单一炊具的性能表现为基础。UL列表根据实验室测试确定排风罩的最低效率。

另一个美国标准，《ASTM标准F-1704-1999——商用厨房通风系统性能测试方法》覆盖了排风罩的吸收和控制性能以及从炊具获得的热增量。现行的ASTM F-1704不设计动态条件，但是正考虑添加能够量化安全系数的动态测试。UL 710和ASTM F-1704对排风罩吸收和控制性能的测试基本相同。

表3.8所显示的排风率是未认证排风罩的最低强制效率，而表3.9则反映了典型的认证排风罩的设计排风率的范围。本表中的数值可用于预估特定项目中认证排风罩与非认证排风罩的风量差别。但是在设计的最后阶段，排风率可能会根据以下条件进行调整：

Chapter 3: Indoor Air Climate in Open Kitchen Restaurants

While the exhaust rates shown in Table 3.8 are minimum mandatory rates for unlisted hoods, the rates in Table 3.9 reflect the typical range in design exhaust rates for listed hoods. The values in this table may be useful for estimating the "cfm" advantage offered by listed hoods over unlisted hoods for a given project. But in the final stage of design, exhaust rates may be adjusted to account for:

1. Diversity of operations (how many of the appliances will be on at the same time).
2. Position under the hood (appliances with strong thermal plumes, located at the end of a hood, tend to spill effluent more easily than the same appliance located in the middle of the hood).
3. Hood overhang (in combination with appliance push-back). Positioning a wall-mounted canopy hood over an appliance line with an 18-inch overhang can dramatically reduce the required ventilation rate when compared to the minimum overhang requirement of 6 inches. Some manufacturers "list" their hoods for a minimum 12-inch overhang, providing an immediate advantage over unlisted hoods.
4. Appliance operating temperature (e.g. a griddle used exclusively by a multi-unit restaurant at 325°F vs. 400°F surface temperature) or other specifics of appliance design (e.g. 18-inch vs. 24-inch deep griddle surface).
5. Differences in effluent from menu selections, such as cooking hamburger on a griddle versus on a charboiler, or using a charboiler to cook chicken versus hamburger.
6. Operating experience of a multi-unit restaurant can be factored into the equation. For example, the CKV system design exhaust rate (for the next new restaurant) may be increased or decreased based on real-world assessments of the CKV system in recently constructed facilities.

1. 操作的差异（同时运行的炊具数量）。
2. 排风罩下方的位置（如果将释放强热卷流的炊具设在排风罩的一端，可能会比放在排风罩中央更易泄漏废气）。
3. 排风罩顶罩（与炊具的后撤式设计相结合）。与6英寸的顶罩相比，在壁挂式伞形排风罩所对应的炊具线上配置一个18英寸的顶罩可以大幅度减少必需的通风率。一些认证制造商保证其所生产的排风罩拥有至少12英寸的顶罩，明显优于未认证的排风罩。
4. 炊具工作温度（例如，在复合餐厅中，325°F的平底锅表面温度对比400°F的平底锅表面温度）或炊具其他细节设计（例如，18英寸的平底锅深度对比24英寸的平底锅深度）。
5. 不同菜品所产生的废气差异。例如在平底锅烹饪牛肉饼对比在烧烤盘上烹饪牛肉饼，或是在烧烤盘上烹饪鸡肉对比烹饪牛肉饼。
6. 复合餐厅的运营经验。例如，商用厨房通风系统的设计排风率可能会随着近期餐饮机构对商用厨房通风系统的工作评价而增减。

Table 3.9 Typical Exhaust Rates for Listed Hoods
表3.9 认证排风罩的典型排风率

Minimum Exhaust Flow Rate for Listed Hoods (cfm per linear foot of hood) 认证排风罩的最低排风率（每纵尺的风量）				
Type of Hood 排风罩类型	Light Duty Equipment 低排放设备	Medium Duty Equipment 中排放设备	Heavy Duty Equipment 高排放设备	Extra-Heavy Duty Equipment 超高排放设备
Wall-mounted Canopy 壁挂式伞形	150-200	200-300	200-400	350+
Single Island Canopy 单岛式伞形	250-300	300-400	300-600	550+
Double Island Canopy 双岛式伞形	150-200	200-300	250-400	500+
Eye Brow 齐眉式	150-250	150-250	not recommended 不推荐	not recommended 不推荐
Back-shelf/Passover 后架式/超越式	100-200	200-300	300-400	not recommended 不推荐

3.4 Ventilated Ceiling

General

The ventilated ceiling is an alternative kitchen exhaust system. The ceiling should be used for aesthetic reasons when open space is required, multiple kitchen equipment of different types is installed and the kitchen floor space is large.

The ventilated ceilings are used in Europe especially in institutional kitchens like schools and hospitals.

3.4 通风顶棚

概述

通风顶棚是另一种厨房通风系统。顶棚的设置通常出于美观原因，主要考虑到开放空间的需要、多种不同类型的厨房设备的安装以及较大的厨房占地面积。

通风顶棚在欧洲，特别是学校和医院的厨房中十分常见。

Chapter 3: Indoor Air Climate in Open Kitchen Restaurants

Ceilings are categorised as "Open" and "Closed" ceiling system.

3.4.1 Open Ceiling
Principle
Open ceiling is the design with suspended ceiling that consists of a supply and exhaust area.

Supply and exhaust air ductworks are connected to the voids above the suspended ceiling. Open ceiling is usually assembled from exhaust and supply cassettes.

The space between the ceiling and the void is used as a plenum. The contaminated air goes via the slot where grease and particles are separated. (See Figure 3.13)

Figure 3.13: Open ceiling
图3.13：开放式顶棚

Specific Advantages
· Good aesthetics.
· Possibility to change kitchen layout.

Disadvantages
· Not recommended for heavy load (gas griddle, broiler).

通风顶棚分为开放式顶棚和封闭式顶棚两种。

3.4.1 开放式顶棚
准则
开放式顶棚配有吊顶，由送风和排风区组成。送风和排风管道与吊顶上方的空洞相连。开放式顶棚通常由送风和排风暗盒装配而成。

顶棚和空洞之间的空间用作增压区。污浊空气先通过条缝分离油脂和烟雾颗粒，然后进入增压区。（见图3.13）

优点
· 美观
· 便于改变厨房布局

缺点
· 不适合高排放型炊具（如燃气平底锅、烧烤架等）
· 只对蒸汽有效
· 不适合卫生要求高的空间（顶棚上方的空间易污染）
· 维护费用高
· 易造成冷凝

第三章　开放式餐厅的室内空气环境

Figure 3.14: Closed ceiling
图3.14：封闭式顶棚

· Efficient when only steam is produced.
· Not recommended from a hygienic point of view (free space above the ceiling used as plenum – risk of contamination).
· Expensive in maintenance.
· Condensation risk.

3.4.2 Closed Ceiling (See Figure 3.14)

Halton ventilated ceiling is based on Capture Jet™ installed flush to the ceiling surface, which helps to guide the heat and impurities towards the extract sections. Supply air is delivered into the kitchen through a low velocity unit.

Air distribution significantly affects thermal comfort and the indoor air quality in the kitchen. There are also combinations of hoods and ventilated ceilings. Heavy frying operations with intensive grease emission are considered to be a problem for ventilated ceilings, so hoods are recommended instead.

Principle
Supply and exhaust units are connected straight to the ductwork. This system consists of having rows of filter and supply units; the rest is covered with infill panel.

There are various closed ceilings. For example, Halton utilise the most efficient ceiling, which includes an exhaust equipped with a high efficiency KSA filter, supply air unit and a Capture JetTM system installed flush to the ceiling panels.

Specific Advantages
· Draught free air distribution into the working zone. from low velocity ceiling mounted panels.
· High efficiency grease filtration using Halton KSA "Multi-cyclone" filters.
· Protection of the building structure from grease, humidity and impurities.
· Modular construction simplifies design, installation and maintenance.
· Integrated Capture Jets within supply air sections.

3.4.2 封闭式顶棚（见图3.14）
封闭式通风顶棚以与天花板平齐的吸收喷射器为基础。喷射器引导热气和杂质进入排风区。不及空气通过低速机组进入厨房。

空气配送影响着厨房内的热舒适度和室内空气质量。排风罩和通风顶棚的结合十分有效。通风顶棚可能无法处理油炸过程产生的大量油烟，而排风罩则能解决这一问题。

准则
送风和排风机组直接与管道相连。这一系统由若干排过滤器送风机组组成。其他部分由填充板覆盖。

封闭式顶棚有多种类型。例如，Halton采用最高效的天花板，其中包括配有高效KAS过滤器的排风系统、送风机组和与天花板平齐的吸收喷射器系统。

优点
· 低速天花板系统将空气配送至工作区
· 高效的油脂过滤系统
· 保护建筑结构不受油脂、湿气和杂质污染
· 部件结构便于设计、安装和维护
· 送气区设有整合式吸收喷射器

Chapter 4: Case Studies

第四章：案例分析

An Open Kitchen Is to Entertain Diners

开放式厨房的作用是取悦就餐者

Muvenpick

莫凡彼餐厅

Restaurant Kaskada

卡斯卡达餐厅

Ellas Melathron Aromata

梅斯拉多兰芳香餐厅

NAMUS Buffet Restaurant

纳姆斯自助餐厅

ZOZOBRA Noodle Bar

祖祖布拉快餐店

Kinugawa

鬼怒川餐厅

Vintaged Restaurant at Hilton Brisbane

布里斯班希尔顿酒店复古餐厅

Lluçanès Restaurant in Barcelona

巴塞罗那卢卡奈斯餐厅

Hilton Pattaya – Restaurants, Lobby & Bar and Linkage Spaces

芭堤雅希尔顿酒店餐厅、大堂、酒吧和连接空间

Assaggio Trattoria Italiana

阿萨吉欧意大利餐厅

Sugarcane

甘蔗餐厅

Symphony's

交响乐餐厅

De Gusto

好味道品鉴餐厅

Karls Kitchen

卡尔斯小厨

The River Café

河畔咖啡馆

Bar Brasserie Restaurant Fitch & Shui

费奇 & 舒伊酒吧餐厅

FRESH CUTT

新鲜出炉烧烤店

Fusao Restaurant

房雄餐厅

La Oliva

奥利瓦餐厅

Grand Hyatt Macau – Restaurant "MEZZA9"

澳门君悦酒店梅萨餐厅

barQue

巴Q餐厅

La Nonna Restaurant

拉诺那餐厅

Jaffa – Tel Aviv Restaurant

特拉维夫雅法餐厅

Aoyama Hyotei

青山冰帝餐厅

Esquire

君子餐厅

Chapter 4: Case Studies

An Open Kitchen Is to Entertain Diners

–Words from designer of Studio Arthur Casas in restaurant design

Sao Paulo is a City that reveals itself behind the walls which makes some places amazing and special. I have noticed a pleasant change in the restaurants of the hotels, especially those completed in the last five years. These places have their own personalities, many times even completely different from the hotels where they are in, with independent entrances when it's possible, and personally I don't see many substantial differences between freestanding restaurants and hotels restaurants.

In my opinion it's absolutely unnecessary to decor restaurants and shops with arts and regarding the colours and materials. I prefer the warm ones (red, orange) and natural materials like stone, brick, wood. They refer to the "affective memory".

开放式厨房的作用是取悦就餐者

——文字来自 Arthur Casas 工作室的餐厅设计师

圣保罗这座城市拥有一些独特而令人惊艳的地点。我注意到酒店餐厅（特别是近五年完成的酒店餐厅）有一些令人愉悦的变化。这些空间拥有自己的个性，一些餐厅甚至与其所在的酒店风格迥异，拥有独立的入口。事实上，我并不认为独立餐厅与酒店餐厅有什么本质上的差别。

我认为完全无需以艺术品以及纷繁的色彩和材料对餐厅和商店进行装饰。我偏爱暖色调（红色、橙色）和自然材料（石材、砖、木材），因为它们与"情绪记忆"相联系。

The discrete façade of the newest restaurant KAA gives no hint of what may prove to be internally. Inside, the narrow space of 798 m² gets a new depth through the Green vertical wall with plants from the Atlantic forest. The water mirror on the bottom of the tropical green wall refers to "Igarapés", that were so common before as well as nowadays in that region.

人们完全无法从 KAA 餐厅的外观想象出其内部的设计。垂直绿化墙为这个 789 平方米的狭长空间赋予了新的深度，墙面上栽种着来自大西洋沿岸森林的植物。热带绿化墙底部的水镜暗指巴西常见的雨林河流。

Chapter 4: Case Studies

The huge stand at the bar that divides the big environment of the restaurant in 2 separate areas is to support indigenous original pieces which mimic with the bottles, cups and books. The roof made with canvas opens automatically. The furniture is contemporary and the philosophy of this place is transporting the urban "Paulista" to a green environment. It's an escape from the chaos.

吧台内的立柜将餐厅分为了两个独立的空间。立柜上摆放的酒瓶、杯子和书籍营造出一种巴西特有的氛围，帆布屋顶可以自动开合。餐厅内家具的设计十分现代，整个餐厅的设计理念是将城市人带到绿洲之中，远离喧嚣。

人们到餐厅不仅仅是为了用餐，更是为了享乐、取悦自己、远离自己的日常工作环境和居所。人们越来越关心食物的来源和它们的做法。餐厅的光线必须适宜，空间不能太大；邻座不能听见自己的对话；音乐必须适当；音响要完美；舒适的桌椅要保证正确的高度。以我来讲，我喜欢看厨师工作，还想自己挑选酒窖中的美酒。

我不迷信技术，并且坚信人们到餐厅和酒吧中不是来寻求高科技的。技术的融合必须自然而和谐。我认为，在餐厅设计中，人们所追求的是一种"情绪记忆"，他们想要找到祖母家、花园里或母亲厨房的感觉。

（所有有关 KAA 餐厅的图片均由 Arthur Casas 工作室设计，Leonardo Finotti 拍摄）

People go to restaurants not only for food but also for entertainment, to surprise themselves, out of their normal working environment and their residences. More and more people's concerns regard to the origin of the food and how it is made. The light must be the right one; the place can't be or looks too large; the neighbour can't hear your conversation; the music must be appropriate; the sound impeccable; a comfortable chair and the table on the correct height. I prefer to see the chef working and if possible choose my wine in the cellar.

I'm not fanatic for technology and I don't believe that people seek for technology when they are in bars and restaurants. If there are technologies it should be in a natural and harmonious way. I really think, especially when we dealing with a restaurant, what people seek is for a kind of "affective memory" of grandmother's house, of a garden, of a mom's kitchen.

(All images are of KAA in São Paulo designed by Studio Arthur Casas and photographed by Leonardo Finotti)

The Drama of an Open Kitchen

Wolfgang Puck practically invented the phenomenon of cooking in front of diners in 1982, Michael Ruhlman wrote in his book, "Reach of a Chef" – it is a claim that discounts Japanese sushi chefs but whatever. What's indisputable is that foodies like to watch food being made.

开放式厨房的魅力

1982年，Wolfgang Puck 创造性地将烹饪过程展示在就餐者面前。Michael Ruhlman 在《厨师的能力》一书中写道：这虽然不太尊重日本寿司师傅，但是美食家们就是愿意看见食物的制作过程。

Chapter 4: Case Studies

Feedback from the Super diners, which San Diego open-kitchen restaurant does good theater and good plates?

The chefs they are culinary artists: cultured, creative and graceful.
— Susan Russo, cookbook author, blogger (foodblogga.blogspot.com)

This year, it has been Searsucker, where Brian Malarkey's "Top Chef" personality comes through. Dine at the kitchen bar and observe the conductor of the food opera. (Editor's note: He's conducting up at his new restaurant Burlap, too.)
— Randee Stratton, real estate broker, avid diner

对话超级食客：圣地亚哥哪家开放式餐厅的表演和美食最好？

厨师就是烹饪艺术家：文明、创新、优雅。
——Susan Russo，食谱作家，博客（foodblogga.blogspot.com）

今年最好的餐厅当属"顶级厨师"Brian Malarkey 的 Searsucker 餐厅。在他的开放式厨房就餐就像观赏一场美食的歌剧。（他也在自己新开的 Burlap 餐厅表演。）

I hate to brag, but I opened one of the first open-kitchen restaurants in town. It is still here and it still serves food (and sushi, too): Sally's Seafood on the Water at the Manchester Grand Hyatt downtown. Chef Sarah Linkenheil serves great food right on the marina, and, as a matter of fact, you can even eat in the kitchen. (But can you stand the heat?) My other favorite is Delicias (6106 Paseo Delicias, Rancho Santa Fe. (858) 756-8000; deliciasrestaurant.com), also one of the oldest restaurant in town with an open kitchen. And guess who was involved with Delicias 20 years ago? Wolfgang Puck! (Editors note: Puck helped open the original Delicias.)

— Fabrice Poigin, private chef, restaurant consultant

Easy Super Cocina (3627 University Ave., Normal Heights. (619) 584-6244; supercocinasd.com). Although it's a cafeteria setup, anyone in line can see the prep and kitchen area clearly. If you're not familiar with a particular guisado (a stew) or just want to taste, simply ask for a sample. That's all the theatre I need.

— Matthew Rowley, food historian, blogger (whiskeyforge.com; Twitter @mbrowley)

I don't know about good theatre, but the open kitchen at Jsix is pretty cool. They recently remodeled their dining room. We had the Chef's Mercy menu there a few weeks ago. Chef Christian Graves cooks up some good meat.

— Robin Taylor, organic farmer at Suzie's Farm and Sun Grown Organic Distributors

When I am paying attention, Dolce Pane e Vino (16081 San Dieguito Road, Rancho Santa Fe. (858) 832-1518; dolcepaneevino.com) seems to keep me looking. I enjoy watching the harmony with which all the cooks in the tiny kitchen are able to produce such good food and not trip all over each other.

— Dave Morgan, CPA, avid diner

我不爱吹嘘，但是我的餐厅是镇上第一家开放式餐厅——Sally's海鲜馆。现在它仍在营业。厨师Sarah Linkenheil在码头制作美食，你甚至可以到厨房进餐（只要你忍受得了厨房的热度）。我的另一家餐厅是Delicias餐厅（6106 Paseo Delicias, Rancho Santa Fe. (858) 756-8000; deliciasrestaurant.com），它也是镇上最早的开放式餐厅之一。而Wolfgang Puck在20年前也参与了Delicias餐厅的筹建。

——Fabrice Poigin，私人厨师，餐厅顾问

Easy Super Cocina餐厅（3627 University Ave., Normal Heights. (619) 584-6244; supercocinasd.com）。尽管这是家自助餐厅，但是人们可以清楚地看到厨房和备餐区。如果你不太熟悉某种炖肉或是想尝一尝，可以直接要求试吃。

——Matthew Rowley，食品历史学家，博客（whiskeyforge.com; Twitter @mbrowley）

我不懂表演，但是Jsix的开放式厨房真是好。他们最近改造了餐室，几个星期前就给我们发了新菜单。主厨Christian Graves做的肉很好吃。

——Robin Taylor，Suzie有机农场的农夫，有机食品经销商

Dolce Pane e Vino餐厅（16081 San Dieguito Road, Rancho Santa Fe. (858) 832-1518; dolcepaneevino.com）一直都引人注目。我喜欢看见厨师们在小厨房里一起制作美食，他们按部就班，不慌不忙。

——Dave Morgan，注册会计师，食客

Chapter 4: Case Studies

Although not classically an open kitchen, there is action behind the counter in the staging area at Kitchen 1540 at L'Auberge Del Mar. Watch as Chef Paul McCabe plates and serves some of the best local ingredients and most original food in San Diego.
— Charles Kaufman, baker, owner of Bread & Cie Café

Having never worked in a restaurant, I am fascinated with all kitchen operations, because it is so foreign to me. The most entertaining kitchen I have seen recently is at Searsucker on a busy night. There is nothing like watching Brian Malarkey yell out orders. The other night I found myself captivated, ignoring my dinner guests, watching the kitchen work like a well-oiled machine. I never did figure out who was listening to Brian's orders, but everything seemed to be working flawlessly.
— David Salisbury, a law firm's director of business development, avid diner

My first Searsucker dining experience started off with us showing up without reservations. The only seating available was at the kitchen counter in full view of the sous chefs and pastry chef working their magic in the open kitchen. They were very personable and to see them prep their amazing dishes was quite the treat.
— Joanne Arellano, human resources professional, avid Yelper, avid diner

Kitchen 1540: I like the setup, it fits the room perfectly.
— William Bradley, executive chef at Addison

I have seen the likes of Emeril Lagasse and Bobby Flay in action, which was a treat. But the best interloping I have been party to involves aproned guys whose names I do not know, whom I have glimpsed when they were not looking. That said, it is fun to watch Brian Malarkey holler to his kitchen commandos at Searsucker. I like getting a peek of the linemen at L'Auberge Del Mar's KITCHEN 1540. And I, like most local foodies, cannot wait to see Matt Gordon's crew at work from the special table that'll be built at Encinitas' upcoming Solace and the Moonlight Lounge.
— Brandon Hernández, food writer (Twitter @offdutyfoodie)

L'Auberge Del Mar 的 1540 厨房并不是一个典型的开放式厨房，但是它就像一个舞台。主厨 Paul McCabe 亲自为人们准备正宗的圣地亚哥美食。
——Charles Kaufman，面包师，经营 Bread & Cie 咖啡馆

因为从没在餐厅工作过，我一直对厨房的运作很感兴趣。一天晚上，我在 Searsucker 餐厅看到了精彩的厨房表演。那天晚上，我对开放式厨房简直着了迷，连宴请的客人都疏忽了。在主厨 Brian Malarkey 的指挥下，一切都井然有序。
——David Salisbury，法律事务所的商业开发部总监，食客

我在 Searsucker 的第一餐并没有预定位子，唯一的空位就是在厨房吧台旁，可以直接看到主厨和面点师傅的精彩表现。他们的表演极好，看着他们准备美食简直是一种享受。
——Joanne Arellano，人力资源人士，食客

1540 厨房。我喜欢它的布局，与整个房间十分契合。
——William Bradley，Addison 餐厅的行政主厨

Emeril Lagasse 和 Bobby Flay 制作美食的过程简直是艺术表演。能看到一群厨师在厨房里忙忙碌碌，实在令人高兴。尤其是看到 Brian Malarkey 在 Searsucker 餐厅的厨房里对手下大呼小叫。我也喜欢看 1540 厨房里的传菜员。我热衷本地美食，期待着看到 Matt Gordon 在 Solace 餐厅和 Moonlight 酒吧的厨艺表演。
——Brandon Hernández，美食评论家（Twitter @offdutyfoodie）

开放式厨房相当有趣。这种路边摊美食所带来的感官刺激也是一种体验。炭火和美食的香气，孩子们

Our farmer's markets offer all kinds of tent-kitchen-theare excitement! The plethora of sensory stimuli that comes with this kind of picnic-nosh-meal makes the experience, I think. The smells of a grill burning and all the different foods cooking; the sounds of children laughing and running around; the feeling of being outside with the sun kissing skin, all the visual delights, and of course, the delicious. Good theatre, good plates! Searsucker, the pizza station at Cucina Urbana and Bread and Cie, too!

— Tina Luu, pastry chef, food lecturer at the Art Institute of California San Diego

When I was little, I was fascinated by flaming tableside preparations of beef and duck at Chinese restaurants. How I did not become a pyromaniac, I will never know! But these days, I keep it simple: barbacoa burrito at Chipotle. It is so exciting how they spread that medium-heat salsa and sprinkle that queso.

— Gerald "Dex" Poindexter, publicist, avid diner

Kitchen 1540: Cannot wait to see where Chef Paul McCabe takes us on his next culinary ride.

— Andrew Spurgin, chef/partner Campine – A Culinary + Cocktail Conspiracy

I gravitate to exhibition kitchens and chef's counters in every city I visit. On our way to Tuscaloosa for a reunion? You can bet we'll be at Hot & Hot Fish Club in Birmingham to watch Chris Hastings at work. But we don't have a favorite here, at least since that once upon a time when Nathan Coulon starred at Quarter Kitchen. Don't tell Brian, but I still haven't eaten at Searsucker, though I hear good things. Does Kono's Café count? Because watching that crew move the number of eggs and trays of bacon that they do through that tiny kitchen is poetry in motion.

— Catt White, farmers' market manager (Twitter @LIMercato)

跑跳欢笑的声音以及阳光下的舒适感都让人心情愉悦。既有好表演，又有美食！Searsucker 餐厅、Cucina Urbana 餐厅的比萨吧和 Bread and Cie 咖啡馆都让人流连忘返。

——Tina Luu，面点师，加州艺术学院圣地亚哥分院的美食讲师

小时候，我一直迷恋中餐馆所展示的烤牛肉和烤鸭。我喜欢燃烧的火焰。现在，Chipotle 餐厅的烧烤让我实现了这个梦想。看着他们在火上翻烤真让人兴奋。

——Gerald "Dex" Poindexter，宣传员，食客

1540 厨房。我都要等不及观看 Paul McCabe 厨师的精彩表演了。

——Andrew Spurgin，Campine 餐厅主厨兼合伙人

我喜欢去开放式餐厅就餐。在前往塔斯卡卢瑟的路上，我必定要去伯明翰的 Hot & Hot 海鲜馆观看 Chris Hastings 的厨艺表演。Quarter 小厨也是我的最爱。我还没去过 Searsucker 餐厅，但是听说那儿很好。Kono's 咖啡馆也不错，员工们在小厨房里端着鸡蛋和培根托盘走来走去，充满了诗意。

——Catt White，农贸市场经理（Twitter @LIMercato）

Muvenpick
莫凡彼餐厅

Completion date: 2009
Location: Zurich, Switzerland
Designer: Stephen Williams Associates
Photos: Mövenpick Hotel Zürich-Airport
Area: 300m²

完成时间：2009年
项目地点：瑞士，苏黎世
设计师：斯蒂芬·威廉姆斯事务所
图片版权：苏黎世机场莫凡彼酒店
面积：300平方米

September 2009 marked the opening of the Mövenpick Restaurant in the Mövenpick Hotel Zürich Airport. Earlier, in February 2009 architects at Stephen Williams Associates, Hamburg, were chosen to take on the project. After five months of planning and advertising and a 20-day construction period, SWA had the restaurant ready for occupancy and could open the doors for business. These establishments were produced, supplied and installed by the international restaurant-fitter Reinhold Keller, Kleinheubach, whose enterprise has already established more than 3,000 restaurants world-wide in just the past few years.

Mövenpick's President and CEO, Jean Gabriel Peres, congratulates with the words: "This is another showcase which sets a trend for future concepts: warm, smart-casual, interactive but quality oriented, I really like it but more importantly the guest feedback is very positive."

2009年9月，苏黎世机场莫凡彼酒店内的莫凡彼餐厅正式开张。斯蒂芬·威廉姆斯事务所于同年2月受委托对餐厅进行设计。经过5个月的规划、宣传和20天的施工，设计师完成了项目。这些成就全部来源于国际餐厅设计师莱茵霍尔德·凯勒——他与他的公司已经在全球打造了3,000多家餐厅。

莫凡彼酒店集团的主席和首席执行官珍·加布里埃尔·佩雷斯对餐厅致以贺词："项目奠定了餐厅设计的潮流趋势：温馨、休闲、智能、互动、高品质。我十分喜欢。重要的是，顾客的反馈也相当好。"

The open kitchen is located at the entrance and diners can appreciate it once stepping into the restaurant.

开放式厨房位于餐厅的入口,就餐者一走进餐厅就能看到它。

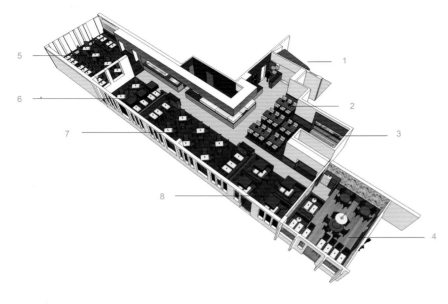

1. Entrance
2. High table
3. Lounge
4. Dim sum
5. Pavilion
6. Separee
7. Flexibel
8. Separee

1. 入口
2. 高桌就餐区
3. 休息区
4. 零食区
5. 大厅就餐区
6. 雅座
7. 灵活就餐区
8. 雅座

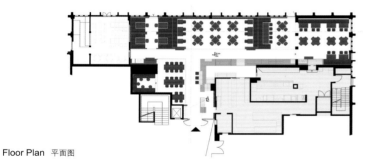

Floor Plan 平面图

The approximate 300m² large surface is separated into six parts with different seat typologies: from low lounge furniture to flexible two- and four-seater tables through to six-seater seating areas and high-tables with bar stools. Each seating combination is suitable for any situation: whether for breakfast, lunch, coffee or dinner, for groups, pairs or those visiting alone – every seat conveys the feeling of being the best. Standing centrally is an L-shaped counter created out of copper. It is here that the meals are prepared in copper pots and pans directly in front of the eyes of the guests. The food is presented on a counter with a copper-tiled base situated below a lintel which is plastered with a coppery metal coating.

Warm walnut, red- and beige-coloured upholstery together with glass panels inlaid with grass help to enhance the freshness and quality of the gastronomy on offer at Mövenpick. Historical black and white photographs of the Zurich-born aerial-view photographer Walter Mittelholzer remind the guest of the wonders of aviation and travel. It is not only the hotel guests, but also the regular Mövenpick customers who are impressed by the new interior design.

近300平方米的平面空间被分为六个区域，配备不同的座椅类型：从低矮的休闲家具到灵活的双人桌、四人桌、六人桌，再到吧台前的高脚凳。每种座椅都适合各种情景：无论是早餐、午餐、咖啡，还是晚餐；无论是团体、双人还是单人。每张座椅都让人感到舒适无比。空间的中央是一个L形的铜制台面。饭菜就在宾客眼前的铜锅中进行烹饪。美食被盛放在铜制台面上，台面上方的横梁同样涂有古铜色金属包层。

温馨的胡桃木、红色和米黄色的软垫与嵌有青草的玻璃板共同凸显了莫凡彼餐厅新鲜的高品质美食。生于苏黎世的著名航拍摄影师沃尔特·米特霍尔泽所拍摄的黑白老照片向宾客们展示着航行和旅行的魅力。无论是酒店房客还是莫凡彼餐厅的常客都对全新的室内设计印象深刻。

Restaurant Kaskada
卡斯卡达餐厅

Completion date: 2010
Location: Liberec, Czech Republic
Designer: maura | Markéta Veselá, Alena Nováčková
Area: 313m²

完成时间：2010年
项目地点：捷克，利贝雷茨
设计师：maura设计公司
项目面积：313平方米

The new restaurant was designed for the empty un-used space, next to the cinema multi-complex, in the Nisa shopping centre in Liberec. The investor convinced the designers about the concept of cooking in front of the guests. The restaurant is visually separated from the shopping area by a strip of lowered ceiling. The maximum height of the restaurant space is preserved and is enhanced by decorated bespoke panelling.

餐厅坐落在利贝雷茨尼萨购物中心影城的旁边。投资人要求设计师设计一个开放式厨房，为顾客展示烹饪技艺。餐厅通过一条低矮的天花板与购物区隔离开。餐厅空间的最高高度得以保留，并且装饰了定制的镶板。

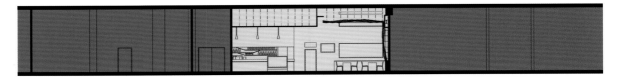

1. Entrance
2. Open kitchen
3. Dining area
4. Bathroom

1. 入口
2. 开放式餐厅
3. 就餐区
4. 洗手间

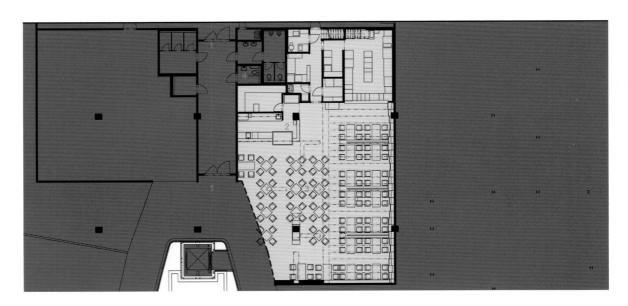

Floor Plan 平面图

The interior of the restaurant is clearly visible from the entrance. The area is divided into three parts: the restaurant, the bar and technical facilities concealed behind a wall on the left side of the restaurant. The cooking area is connected with the restaurant space by a narrow horizontal window, making the whole operation of the kitchen clearly visible and allowing patrons to take a peek at the chefs at work. Diagonal to the kitchen is the dishwashing area and a long corridor with preparation areas and storage. A single stairway leads to the floor with facilities for employees.

从入口处就可以对餐厅的内部一览无余。整个空间被分为三个部分：餐厅、酒吧和隐藏在餐厅左侧墙壁后方的技术设施。烹饪区通过狭窄的横向窗口与餐厅空间相连，将整个厨房清晰地展现在顾客面前。厨房斜对着洗碗区和一条配置着备餐区和仓库的长走廊。狭窄的楼梯连接了上方的员工设施。

The main motif of the wooden panelling is based on the name of the restaurant. The aesthetic panelling of the central column of the restaurant also reflects the motif. Here, the wooden panels on two sides of the column conceal storage. Visitors can instantly see the elegant silhouette of the letter "K", enhanced by illuminated glass, affixed to the column. In addition to wooden panelling, another important material incorporated is roughly stacked sabre slate forming low partitioning walls between the tables, the bar wall and the rear wall of the restaurant. The remaining surfaces of the lowered ceiling are painted in an anthracite tone and the cream coloured walls are livened up with graphic printing again based on the letter "K".

木镶板上的主要图案以餐厅名字中的"K"为基础。餐厅中央的立柱上同样反映了这个图案。柱子两侧的木板将仓库隐藏了起来。由于立柱上安装了发光的玻璃板，顾客可以一眼就看到优雅的"K"字母。除了木镶板之外，另一种重要的材料是餐桌之间由粗面岩板形成的低矮隔断、吧台墙和餐厅后墙。低天花板的表面以炭黑色为主，而乳白色的墙壁则衬托了"K"字印花图案。

Ellas Melathron Aromata

梅斯拉多兰芳香餐厅

Completion Date: 2011
Location: Athens, Greece
Designer: Minas Kosmidis
Photos: Ioanna Roufopoulou
Area: Inside 214m², Balcony 97m², Open Space 190m²

完成时间：2011年
项目地点：希腊，雅典
设计师：米纳斯·科斯米迪斯
图片版权：艾欧亚纳·罗孚普洛
面积：室内214平方米，阳台97平方米，开放空间190平方米

Ellas Melathron Aromata is located at the shopping centre of Smart Park at Spata of Attica, in Greece. The concept of this alternative coffee-candy-fast food shop was the creation of a space with intense scent, flavours and atmosphere of urban Greece in the early 50's and 60's.

The store is developed on three floors. In the ground floor there are the principal services for the public, while the living room expands on the first floor. At the basement there are the ancillary areas, the main kitchen and the toilets, while the store expands with tables and chairs on a covered outdoor patio, which is bordered by vertical wooden beams that form the perimeter and the other sedentary. The pergola has panels fitted with digital prints and rosettes with hanging lights to illuminate the exterior.

梅斯拉多兰芳香餐厅位于希腊阿提卡智能园的购物中心。餐厅主要供应咖啡、甜点和快餐，其设计理念旨在打造一个具有浓郁的20世纪五六十年代希腊城市气息的空间。

餐厅分为三层。一层主要为公众服务，生活空间延伸到二层，地下室则是厨房、洗手间等附属空间。餐厅的露天平台上摆放着桌椅，四周的木梁形成了边界，供人休憩。绿廊的嵌板上配有数码印花，悬挂的吊灯则为室外空间提供照明。

The open kitchen is located on the ground floor, which can be seen from outside through the glass façade as well. The diners can enjoy the skillful performance of the chef while having their delicious food.

Dining areas on the ground floor are separated and furnished in different styles. The intense Greek character is underlined by the traditional minimarket-like booth. Other elements adding to the scenic appearance of the environment are the handmade mosaic tiles on the floor, the wooden surfaces on the walls, the lighting fixtures and the furniture fabrics. In addition, the shade colours of grey, white and blue, as well as the pictures of the Neo-classical era of Greek architecture, complete the puzzle.

开放式厨房位于一楼，过往的行人可以透过玻璃墙看到。就餐者可以在享用美餐的同时观赏大厨的精湛表演。

一楼的就餐区被划分为几个不同风格的独立空间。与传统市场相似的摊位彰显了浓郁的希腊风情。其他装饰元素包括手工制作的马赛克地砖、木制墙面、灯具和家具织物等。此外，灰色、白色和蓝色的叠加以及新古典时期的希腊建筑图画都为空间增色不少。

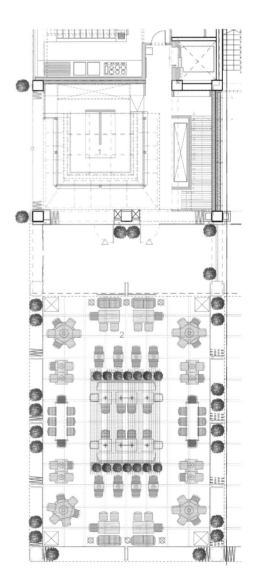

1. Open kitchen
2. Dining area

1. 开放式厨房
2. 就餐区

Ground Floor Plan　一层平面图

Dining area on the first floor brings outdoor scene inside through green plantings implanted in, which also compliments the charming scenery outside and creates a soothing atmosphere for diners. The lighting fixtures dangle from the ceiling and the cable lines serve as decorative elements.

二楼就餐区通过绿色植物将露天景观引入了室内。绿色植物同样美化了室外空间，为顾客营造了舒缓的氛围。从天花板上垂吊下来的灯具和电线起到了装饰作用。

1. Lounge area
2. Dining area
3. Toilet
4. Outdoor dining

1. 休息区
2. 就餐区
3. 洗手间
4. 露天就餐区

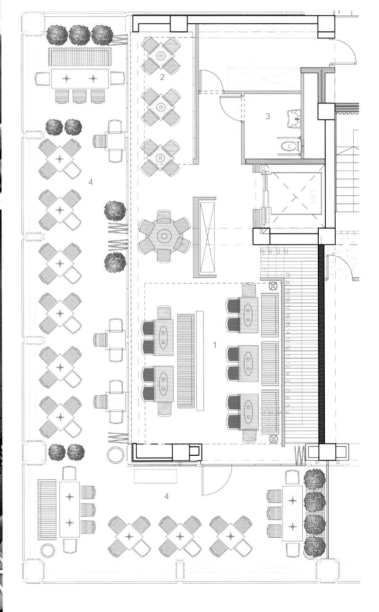

First Floor Plan　二层平面图

NAMUS Buffet Restaurant
纳姆斯自助餐厅

Completion date: 2011
Location: Gyeonggi, Bundang, Korea
Designer: Kim Chiho/Chiho&Partners
Photos: Indiphos
Area: 980m²
Material: (Floor)Tile, Carpet ; (Walls)Wood louvre panel, Special paint, Metal fabric; (Ceiling) Special paint

完成时间：2011年
项目地点：韩国，京畿道，盆唐区
设计师：金志宏/志宏事务所
图片版权：Indiphos
面积：480平方米
材料：（地面）瓷砖、地毯；（墙壁）木制百叶板、特殊涂料、金属网；（天花板）特殊涂料

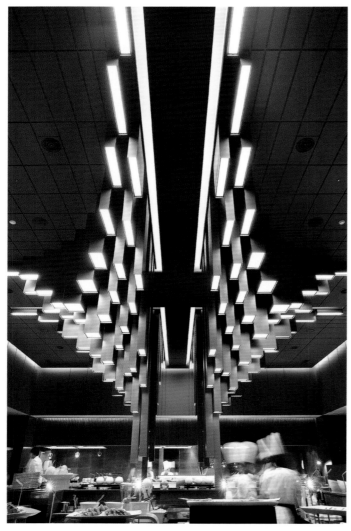

Continuity of geometric structure makes potential space-esthetics by showing various compositions. This project is extracted from a multiple sequence of progressively flattened geometric elements characterised by "continuity and rhythm" blowing through the space, put on the uniform rules.

The restaurant consists of a main hall in centre, 3 more halls and 6 rooms with a unique concept. Using natural materials like stone, wood and steel makes the restaurant environmentally friendly. At the same time it is emphasised its warmth, elegance, colour and texture of the materials.

Centred on a communal table, Kikoski's channeled Lloyd Wright's iconic spiraling white design from 1959 in the nearly all-white 1,600-square-foot-space.

A reasonable lighting plan associated with six different pastel coloured rooms creates a stable illumination and light lines and it makes a comfortable dining environment.

连续的几何结构通过展示各种各样的组合创造了潜在的空间美学。项目由多重平面结合元素序列抽象而成，以连续性和节奏感为特色，形成了统一的规则。

餐厅由一个中心大厅，三个小厅和六间包房组成，拥有独特的主题。石材、木材和钢铁等自然材料的运用使餐厅十分环保，同时也凸显了餐厅的温馨、优雅、色彩和材料的质感。

良好的照明设计与六间各不相同的彩色包房打造了稳定的照明和光线，形成了舒适的就餐环境。

餐厅位于1959年面积约147平方米的近乎纯白空间内，设计以共享桌为中心，把餐厅打造成螺旋形状。

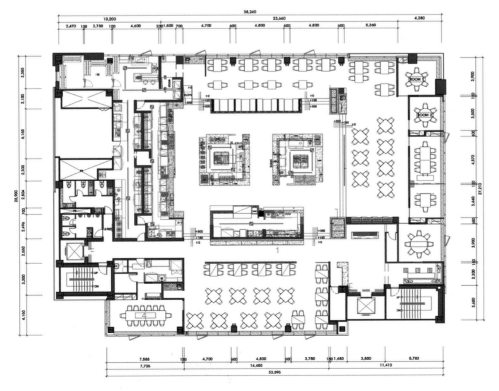

1. Hall
2. Japanese kitchen
3. Cool kitchen
4. VIP room

1. 大厅
2. 日式餐厨房
3. 冷餐厨房
4. 贵宾就餐区

Floor Plan　平面图

The open-kitchen resides in the main hall, as a central stage where the chefs are busy performing their cooking technique. The sculpture hanging from the ceiling in a main hall reminds of the future metropolis image. It is also an icon of the whole space with the symbolism motivated by geometric nature.

开放式厨房设在主厅内，大厨们在这个中心舞台上表演着精湛的厨艺。从天花板上吊下来的雕塑作品模仿了未来都市的形态。它也是整个空间的标志物，体现了几何的本质。

The design concept is reliable and energetic features of absolute beauty in a space. Highlighted lines are constitutive elements that separate finishing materials vertically. Horizontal lines are applied to the element that separates the space and makes a path. The dining areas are of different styles but there is a common connection in colour and material selection.

Private rooms provide an intimate dining experience where diners can enjoy their nice meal in a tranquil as well as cozy space. More attention is paid to the lighting fixtures.

设计理念体现了空间绝对美感的可靠性和活跃性。亮线是垂直分割装饰材料的重要组成元素。水平线条的运用则将各个空间隔开，形成了路径。就餐区的风格各异，但是在色彩和材料选择上具有一致性。

包房提供了私密的就餐体验，就餐者可以在安静舒适的空间内享用美食。包房设计更注重灯具的选择。

ZOZOBRA Noodle Bar
祖祖布拉快餐店

Completion date: 2011
Location: Israel
Architects: Alon Baranowitz and Irene Kronenberg; www.bkarc.com
Client: Ben Rothschild in partnership with chef Avi Conforty
Photos: Amit Geron
Area: 450m²

完成时间：2011年
项目地点：以色列
建筑师：阿龙·巴拉诺维奇与艾琳·克隆恩伯格
委托人：本·罗斯切尔德与阿维·康福迪
图片版权：阿密特·格龙
面积：450平方米

ZOZOBRA is not a romantic space. It is an edgy space, ecstatic and cool.

On the outside, the designers introduced "ZOZOBRA PARK", a linear "hilly" play of topography flanking the restaurant where restless kids can play while their parents enjoy a quiet moment. Planted amongst these hills are two vertical screens communicating ZOZOBRA's world to the street.

ZOZOBRA offers a dynamic eating experience through a fast service supported by an energetic cool staff geared with wireless pads, a sizzling state of the art open kitchen, and a local service code which states: "Whichever dish is ready, we serve to the table. No firsts, no mains…"

祖祖布拉快餐店一点也不浪漫，它前卫、迷人、出色。

设计师在餐厅外部引入了"祖祖布拉公园"——丘陵式的地形空间。顽皮的孩子可以在这里玩耍，让他们的家长静享美食。两块立式屏幕向街道展示了祖祖布拉的世界。

祖祖布拉快餐店的服务十分快捷：充满活力的店员们配有无线设备，开放式厨房永远忙碌。他们遵循当地的服务法则："哪样菜品先做好，就先上菜。无关先来后到，无关身份地位。"

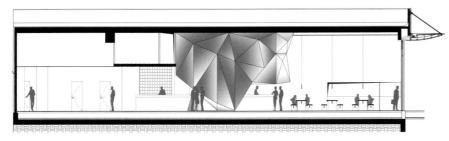
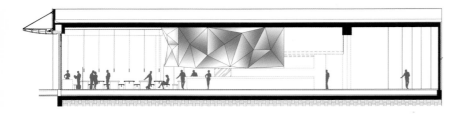
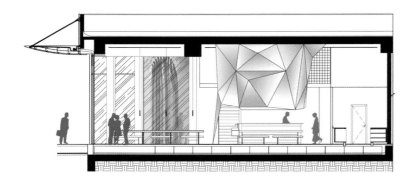

Sections 剖面图

All of these features encourage guests to react to the space and to each other. Technology is a key feature in our lives and so does ZOZOBRA. It reacts to that reality and embraces it.

The open kitchen is a key element of the layout scheme. It is a powerful feature; the heart and soul of ZOZOBRA.

The designers did not settle for just exposing it, they needed to radiate out the energies which burst within its counter tops, works and grills.

A three-dimensional origami structure folds around the kitchen, wrapping up its space like a white cloud, creating a sort of a tempest "dancing" around the kitchen. While looking into the kitchen, not only guests would see 12 people "in battle", but also, a strong movement and vitality played by the architecture to compliment it.

所有特征都鼓励客人与空间或相互之间进行互动。技术是我们生活中的重要元素，也是祖祖布拉所不可或缺的。它与现实互动，并接受现实。

开放式厨房是空间布局的重要元素，它是祖祖布拉的心脏和灵魂所在。

设计师不满足于仅仅展示厨房，他们想要让厨房通过台面、作品和烤架散发能量。

一个立体折纸结构将厨房环绕起来，仿佛白云在围绕着厨房翩翩起舞。人们不仅能看到厨房里12个人的忙碌身影，还能感受到建筑的强烈动感和活力。

Up with the cloud to the gallery level is where the public toilets are located.

The highly reflective black membrane finish of the restaurant ceiling amplifies and reflects anything that happens on the floor.

云朵的上方是公共洗手间所在的夹层楼。

餐厅天花板的黑色高反光薄膜能够放大并反射地面上发生的一切。

1. Service kitchen　　　　　1. 服务厨房
2. Preparation kitchen　　　2. 备餐厨房
3. Dinning　　　　　　　　3. 就餐区
4. Bar　　　　　　　　　　4. 吧台
5. Dish washing area　　　　5. 洗碗间
6. Restrooms　　　　　　　6. 洗手间
7. Food&beverage storage　 7. 食品饮料仓库
8. Take away order room　　8. 打包室
9. Pasta prep area　　　　　9. 意大利面准备区
10. Mechanical room　　　　10. 机房
11. Offices　　　　　　　　11. 办公室
12. Staff changing room　　 12. 员工更衣室
13. General storage　　　　 13. 总储藏室
14. Men's restroom　　　　　14. 男洗手间
15. Qomen's restroom　　　　15. 女洗手间

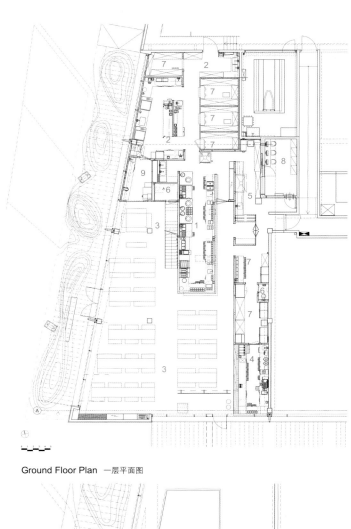

Ground Floor Plan 一层平面图

First Floor Plan 二层平面图

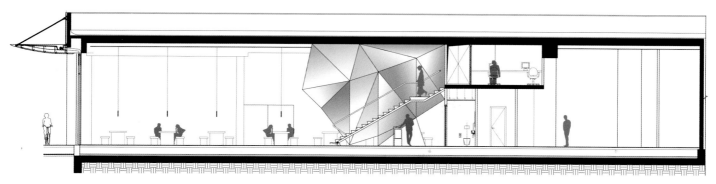

Section 剖面图

At ZOZOBRA, private territory is blurred and replaced by a chance for social interaction. The ambiance of the venue varies as the day progresses thanks to the ever-changing LED lighting schemes and the video art, specially directed for the project, shooting the inner world of ZOZOBRA on the walls.

在祖祖布拉，私人界限被模糊，由社交互动所替代。变幻的LED照明方案和影像艺术让空间的氛围不断变化。视频通过投影，将祖祖布拉的内部世界投射在墙面上。

Kinugawa

鬼怒川餐厅

Completion date: 2012
Location: Paris, France
Designer: Gilles & Boissier
Photos: Matthieu Salvaing

完成时间：2012年
项目地点：法国，巴黎
设计师：吉勒斯&博伊斯尔
图片版权：马修·萨尔文

Design Brief
Helping the renowned restaurant "Kinugawa" – long acknowledged as the Parisian hub of Japanese gourmet cuisine – to further its reputation by creating a space that lends itself to the classic Japanese culinary experience, with echoes of Japanese "Zen" lifestyle. The starting concept was to find a way to combine Japanese culture with French social etiquette and "joie de vivre". Gilles & Boissier were chosen by the restaurant owner to rise to the challenge of creating a space which emanates a jovial & complaisant atmosphere as experienced in a traditional Parisian "brasserie" in an overtly Japanese context.

Design Challenge

The challenge for Gilles & Boissier was bring to life a design which embodies both the traditional & contemporary soul of Japan. The two-sided nature of the Japanese culture is inspired and embodied in Japan's two major cities – Tokyo & Kyoto. The contemporary aspect of the design takes its influence directly from daring & dynamic Tokyo; the traditional aspects explore the tenets of the Zen aesthetic, referring to contemplative & calm Kyoto. The most testing aspect of the project was the amalgamation of both contemporary & traditional Japanese culture with a French touch.

设计概要

设计旨在帮助位于巴黎的著名日本餐厅"鬼怒川"扩大声名，打造了一个融合经典日式烹饪体验和日本"禅"生活的美食空间。设计的出发点是找到日本文化与法国社交礼仪和生活乐趣的结合点。餐厅业主选择了吉勒斯 & 博伊斯尔来实现这一设计，并要求他们打造一个散发出愉悦而谦逊氛围的空间，在日本餐厅的环境中打造法国啤酒屋的氛围。

设计挑战

设计师所面临的挑战是为象征着日本传统和现代之魂的设计注入活力。日本文化的两面性在它的两座城市中得以体现——东京和京都。设计的现代感来自于大胆活跃的东京；传统感则探索了禅宗美学的原理，参考了沉静平和的京都。项目最棘手的部分就是现代和传统日本文化与法国风情的结合。

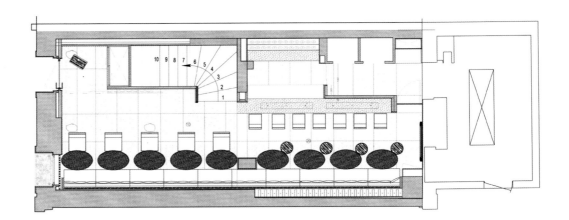

Open Kitchen　开放式厨房区

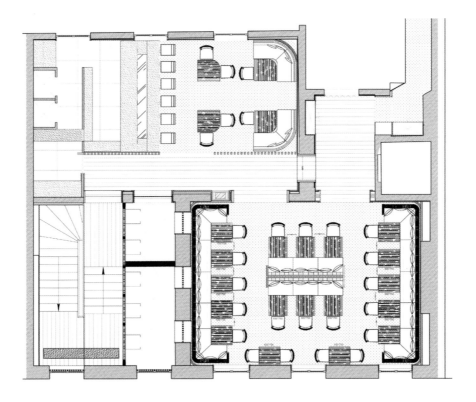

Dinging Area　就餐区

The open-kitchen is considered as the focal point of the restaurant, which creates a stage effect for diners to enjoy the chef's performance while eating their nice meal.

开放式厨房是餐厅的焦点，营造出一种舞台的效果。就餐者可以边观看厨师的表演，边享用美餐。

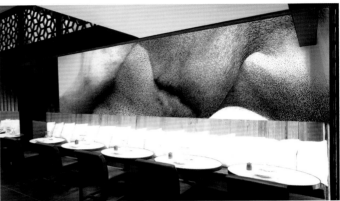

Wooden openwork pillared partitions are used throughout the space and are reflected along the mirrored staircase wall, which plays on a theme of mixing light and shadow as found within Japanese theatre.

The first floor serves as the "workshop". In this large, more traditional space people find an eclectic combination of raw materials used to give depth and texture to the walls. Ranging from darkened cedar wood, scorched stone-work & raw leather seating, such use of natural materials is elementary to all G& B design projects. The floor is dominated by an elegant & stylish mottled-red floor-rug, alluding to the geishas of antiquity.

An impressive 6m high wall mural – the original work of artist Alix Waline – invokes the motion & movement of water and transports people to the Japanese water gardens as depicted in traditional Japanese embossed art. The size and presence of this piece is juxtaposed with its subtle colours and "Zen" undertones in accordance with the overall "Tokyo & Kyoto/Modern & Classic" design theme.

餐厅内部大面积使用了木制透雕隔断，它们经过楼梯间镜子墙面的反射，形成了日本剧院常见的光影混合主题。

二楼被设计成一个"工坊"。这个大型传统空间的设计结合了各种原材料，赋予了墙壁深度和质感。深色杉木、灼烧石制品和生皮革座椅等自然材料的运用是设计师的必备元素。地面上铺设着优雅时尚的红色斑点地毯，暗指古典艺妓的妆容。

高达6米的巨幅壁画（艺术家阿历克斯·瓦林的原创作品）暗指动态的水流，将人们带到了日本传统浮雕艺术中所描绘的日式水景花园。壁画的尺寸与其微妙的色彩和禅意形成了鲜明对比，正好切合"东京与京都，现代与古典"的设计主题。

Vintaged Restaurant at Hilton Brisbane

布里斯班希尔顿酒店复古餐厅

Completion date: 2011
Location: Brisbane, Australia
Designer: Landini Associates Pty Ltd
Photos: Trevor Mein – Mein Photography
Area: 566m² (restaurant & kitchen) + 200m² (bar)

完成时间：2011年
项目地点：澳大利亚，布里斯班
设计师：兰迪尼设计事务所
图片版权：特雷弗·梅恩——梅恩摄影
面积：566平方米（餐厅&厨房）+200平方米（酒吧）

In 1985 the world famous Modernist architect Harry Seidler completed the only hotel he was ever commissioned to design. Twenty-five years later Landini Associates have remodeled its restaurant and bar. Intended in part as "homage" to the great man Landini took their cues from the building's splendid grandeur and have designed a classic space with a timeless simplicity.

1985年，世界著名现代主义建筑师哈里·赛德勒完成了他唯一的一座酒店设计。25年后，兰迪尼设计事务所对酒店的餐厅和酒吧进行了改造。为了向这位伟人致敬，兰迪尼事务所从建筑恢弘的设计中获取了灵感，打造了一个简约恒久的经典空间。

1. Existing counter
2. Existing piano
3. Existing banquette seating to remain
4. Bar
5. Existing banquette seating to remain
6. Mirror
7. Glazing
8. Mobile unit
9. Buffet
10. Grill
11. Cheese table
12. Cold wall
13. Prep cooking
14. Dry age case
15. Bar storage
16. Bar cool room
17. Office
18. Lift
19. Relocated cool room
20. Maitre d'
21. Wine Table
22. Wine Room
23. Female WC
24. Male WC
25. Private dining room
26. Storage
27. Waiter Station
28. Temporary wall & doors to be installed

1. 已有柜台
2. 已有钢琴
3. 已有餐位
4. 酒吧
5. 已有餐位
6. 镜子
7. 玻璃墙
8. 移动单元
9. 自助餐
10. 烤肉区
11. 芝士桌
12. 冷墙
13. 备餐区
14. 干腌柜
15. 酒吧仓库
16. 酒吧冷藏室
17. 办公室
18. 电梯
19. 迁移的冷藏室
20. 餐厅领班
21. 酒桌
22. 品酒室
23. 女洗手间
24. 男洗手间
25. 包房
26. 仓库
27. 服务台
28. 临时墙壁和门

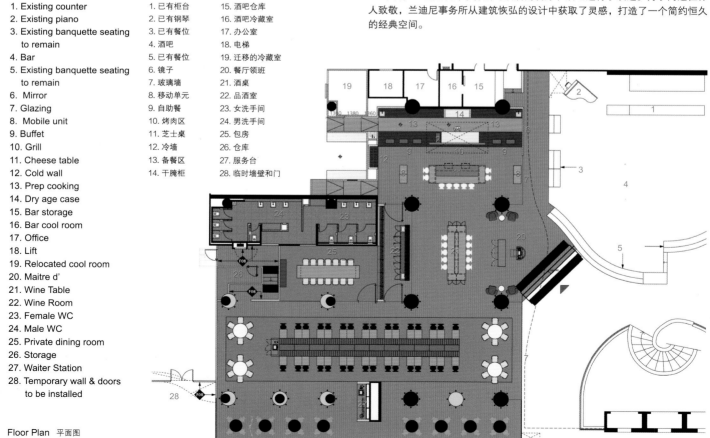

Floor Plan 平面图

The restaurant is located on a raised platform to separate it from the surrounding reception and public spaces. It is entered by a wide stair leading from its bar which sits under the twenty-five floor atrium.

On entering one is greeted by the Maître D who operates from a stainless steel drum, framed by the backdrop of the restaurant's glazed wine room. To the right is an open kitchen and grill, to the left the main dining room. A chef's table sits in front of the open cooking area, which is constructed entirely in black stone to highlight both the aging beef and the chefs who prepare it.

In front of the wine room is a long table for groups and wine tasting classes. These dark finishes extend the length of the dining room and public corridor along which it runs. Mirrors dress this public space and are fitted with vertical lights for added glamour.

餐厅位于抬高的平台上,与周围的酒店前台和公共空间隔离开。中庭高达25层,宾客通过连接中庭酒吧的大楼梯进入餐厅。

一进门,站在不锈钢圆柱旁的餐厅领班将引领宾客进入餐厅。靠近大门的是餐厅的玻璃品酒室。右侧是开放式厨房和烤肉区,左侧则是主餐室。开放式厨房前方的厨师桌全部由黑色岩石制成,突出了烘烤的牛肉和忙碌的厨师。

品酒室前方是一张长桌,供团体品鉴和品酒教学使用。黑色的装饰拓展了餐室和公共走廊的长度。垂直照明为公共空间的镜子增添了光彩。

Dark leather custom designed banquettes, walnut and ebony chairs and tables allow the modernist red lounges to complement the otherwise white interior of the hotel.

Artwork was specially commissioned but also found in drawings unearthed from Seidler's studio and then enlarged and placed behind the monochrome reception.

黑色定制皮革座椅、胡桃木和黑檀木餐桌椅让现代风格的红色休息室为以白色为主的酒店室内设计增添了色彩。

前台背景墙上的艺术品一部分是特别定制的，一部分则由赛德勒工作室的图纸放大而成。

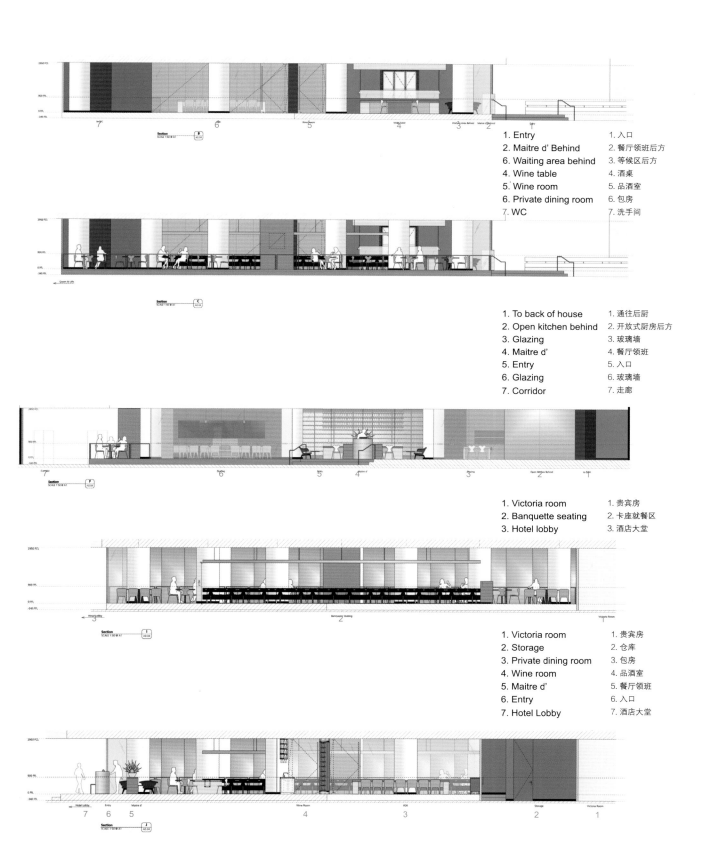

1. Buffet 1. 自助餐区
2. Open grill 2. 开放式烤肉区
3. Buffet 3. 自助餐区
4. Coffee pass 4. 咖啡台
5. Kitchen 5. 厨房

1. Entry 1. 入口
2. Maitre d' Behind 2. 餐厅领班后方
3. Waiting area behind 3. 等候区后方
4. Wine table 4. 酒桌
5. Wine room 5. 品酒室
6. Private dining room 6. 包房
7. WC 7. 洗手间

1. To back of house 1. 通往后厨
2. Open kitchen behind 2. 开放式厨房后方
3. Glazing 3. 玻璃墙
4. Maitre d' 4. 餐厅领班
5. Entry 5. 入口
6. Glazing 6. 玻璃墙
7. Corridor 7. 走廊

1. Victoria room 1. 贵宾房
2. Banquette seating 2. 卡座就餐区
3. Hotel lobby 3. 酒店大堂

1. Victoria room 1. 贵宾房
2. Storage 2. 仓库
3. Private dining room 3. 包房
4. Wine room 4. 品酒室
5. Maitre d' 5. 餐厅领班
6. Entry 6. 入口
7. Hotel Lobby 7. 酒店大堂

123

Lluçanès Restaurant in Barcelona

巴塞罗那卢卡奈斯餐厅

Location: Barcelona, Spain
Designer: Josep Ferrando
Photos: Adria Goula
Area: 500m²

项目地点：西班牙，巴塞罗那
设计师：约瑟夫·弗兰多
图片版权：阿德利亚·古拉
面积：500平方米

The programme is divided between the two floors, on the ground floor a more informal, and at the top, a Michelin Star restaurant.

The architectural solutions to distribute the programme are both light and acoustic response of the space. The lamp is solved by a steel plate painted white and perforated according to a mosaic that remembers your settings so characteristic of the neighbourhood where the restaurant is situated. Behind the plate, perforated fabric absorbs while acoustically diffuses light.

The upper and lower position of the lights generate a dialogue between the static (the project) and dynamic (the user): adjust light intensity when you are standing (dynamic) and light in darkness when you sit (static).

The lower fixture gives lightness and buoyancy to the element aspect lamp located inside the kitchen and bathrooms.

餐厅根据楼层分为两个部分，下层比较休闲，而上层则是米其林星级餐厅。

建筑师通过灯光和音响效果将两个空间区分开来。白漆钢板灯罩上嵌有马赛克式图案，使餐厅布景体现了其所在社区的特色。钢板后方的穿孔织物能吸收并漫射光线。

上下两个方向的灯在静止（项目）和动态（用户）之间形成了对话：使用者站起时将灯光调亮，坐下时则将灯光调暗。

较低的灯具为厨房和洗手间提供了轻快放松的感觉。

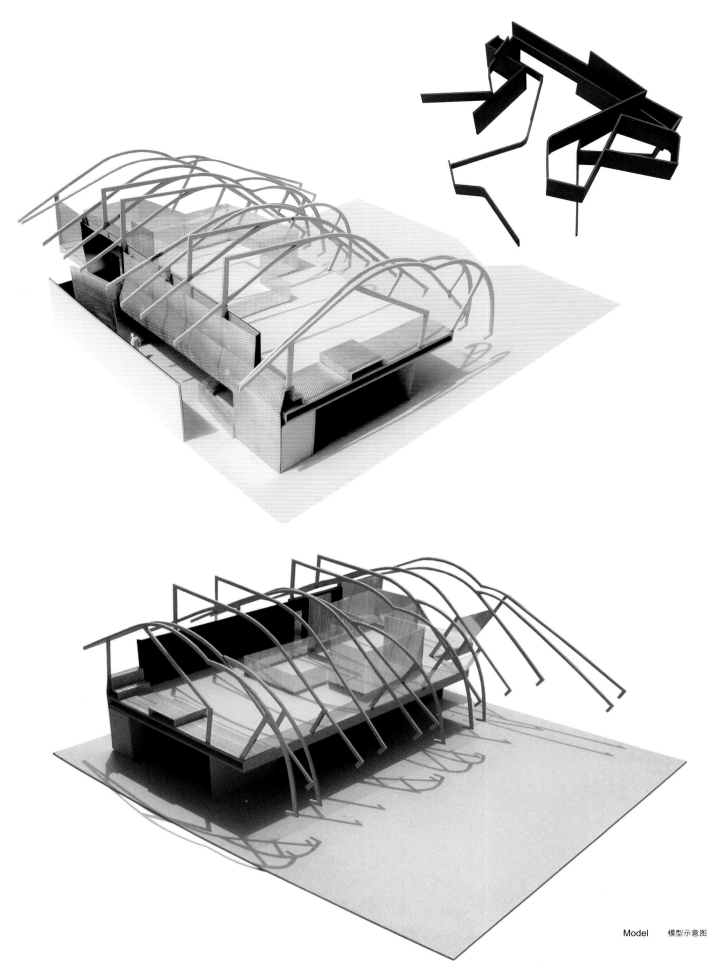

Model 模型示意图

This ground floor space as a continuum of the square, with a large opening in front and a continuous concrete floor, becomes a door space between public and market. The layout creates two working parties on both sides of the flow of people: one open, where the layout of the kitchens in perpendicular promotes cooking show, and one closed to the dirty area with access from the outside service. Lighting along the length of the place reinforces the idea of continuity, turning the square into a restaurant and the restaurant plaza.

一层空间是外面广场的延续，巨大的前门和大面积的水泥地面使其成为了公众与市场之间的门户。空间布局将人流分为两部分：一部分是开放的——垂直的厨房布局让人们可以欣赏烹饪表演；另一部分是不干净的外部服务区。沿着空间直线的灯光突出了连续性设计，让广场变成了餐厅，餐厅变成了广场。

The walls painted black background reinforce the sense of infinity from the public space. On the ground floor facing a large blackboard acts black, while the facing first floor fragments to locate the necessary furniture, leaves two large openings that give visual continuity to the deck in the double-height space. The furniture is situated opposite the skylights to increase light coming downstairs.

以黑色为背景的墙面凸显了公共空间的无尽感。一层墙上设置着一面巨大的黑板，二层分割的墙面上则摆放着必要的家具，两个巨大的开口连接了平台与双层空间。家具的摆放与天窗相对，增加了射向楼下的光线。

The first floor is achieved by a great cabinet space; positioned asymmetrically in the room, it creates four places between the perimeter of the façades and to which never touches, so that the container and metal beams have continuity. The floor, with references to aerial photographs of fields, indicating the importance of raw material at a restaurant in this category, while carpets generates three intensities of gray that individualise each table space, avoiding the feeling of great dining room.

Gray tones on the first floor are repeated in the dress of the tables, as if they were part of the floor, while ground floor provides a single colour, orange, which qualifies the different types of premises. This colour is always at the height of the sight of the person sitting on chairs, stools, aprons. The materiality of the proposal seeks to maintain the industrial character of the market, using steel plates and concrete.

二楼空间由一个大橱柜组织而成。橱柜摆放在房间的对称线上，隔成了四个空间，连接了金属梁和整个空间。地面参考了田地的航拍图，突出了餐厅食品原料的重要性，而三种不同灰度的地毯则区分了各个餐桌空间，避免了大餐厅的感觉。

二楼的灰色色调在餐桌布置上得以重复出现，使其看起来像是地面的一部分，而一楼则以橙色为单一色调，标志了不同的受众。橙色总是出现在人们在座椅和长凳上的视线高度。设计所采用的材料保持了市场的工业特征，以钢板和水泥为主。

The white wall-lamp at an altitude of just 2.8m, from which the black takes away space for installation.

白色壁灯的高度正好为2.8米,黑色从这里开始向上延续。

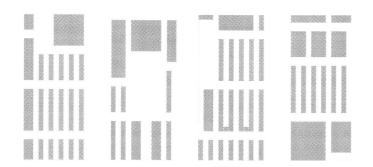

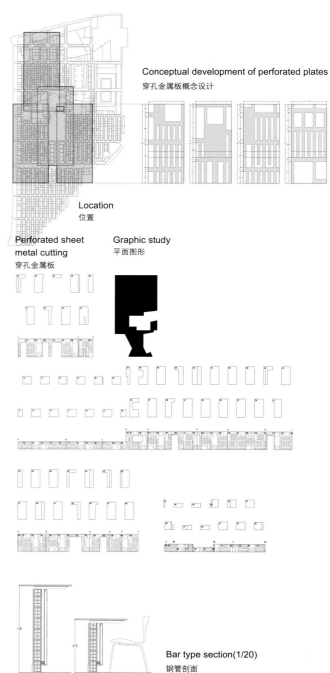

Conceptual development of perforated plates
穿孔金属板概念设计

Location
位置

Perforated sheet metal cutting
穿孔金属板

Graphic study
平面图形

Bar type section(1/20)
钢管剖面

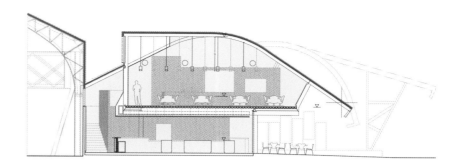

1. Kitchen
2. Refrigerating chamber
3. Preparation area
4. Cleaning area
5. Café area
6. Women's restroom
7. Men's restroom
8. Toilet for handicaps
9. VIP room
10. Dining room

1. 厨房
2. 冷库
3. 备餐区
4. 清洁区
5. 咖啡区
6. 女洗手间
7. 男洗手间
8. 残疾人洗手间
9. 贵宾室
10. 餐厅

1. Cleaning area
2. Kitchen
3. Dining room
4. Refrigerating chamber
5. Stock
6. Machinery
7. Control area
8. Reception material
9. Service access
10. Terrace

1. 清洁区
2. 厨房
3. 餐厅
4. 冷库
5. 仓库
6. 机房
7. 控制区
8. 原料接收
9. 服务入口
10. 平台

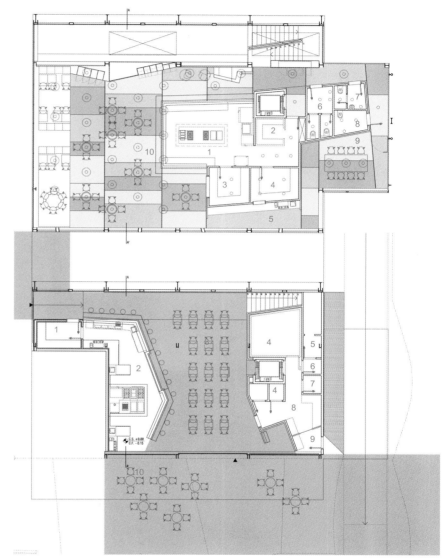

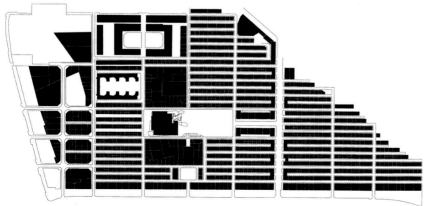

Site Plan 总平面图

Hilton Pattaya – Restaurants, Lobby & Bar and Linkage Spaces

芭堤雅希尔顿酒店餐厅、大堂、酒吧和连接空间

Completion date: 2010
Location: Pattaya, Thailand
Designer: Waraphan Watanakaroon
Photos: Wison Tungthunya
Area: Lobby and "Drift" Bar: 1,650m^2; "Edge" Restaurant: 1,200m^2; "Flare" Fine Dining: 205m^2

完成时间：2010年
项目地点：泰国，芭堤雅
设计师：瓦拉方·瓦塔纳卡隆
图片版权：维森·唐苏雅
面积：大堂和漂流酒吧：1,650平方米；边缘餐厅：1,200平方米；闪耀高级餐厅：205平方米

Waraphan Watanakaroon is responsible for interior design of various common areas for Hilton Pattaya Hotel which includes the Ground Floor Lobby, the Main Lobby on the 16th floor, "Drift" Bar, "Edge" Restaurant, "Flare" Fine Dining, and various common area and linkage spaces within the building. The hotel is part of a larger multi-used complex located in the heart of Pattaya, overlooking the Pattaya beach.

瓦拉方·瓦塔纳卡隆负责对芭堤雅希尔顿酒店的公共空间进行设计，其中包括：一楼大堂、17楼大堂、漂流酒吧、边缘餐厅、闪耀高级餐厅和各种公共空间及连接空间。酒店是芭堤雅中心大型综合体项目的一部分，俯瞰着芭堤雅海滨。

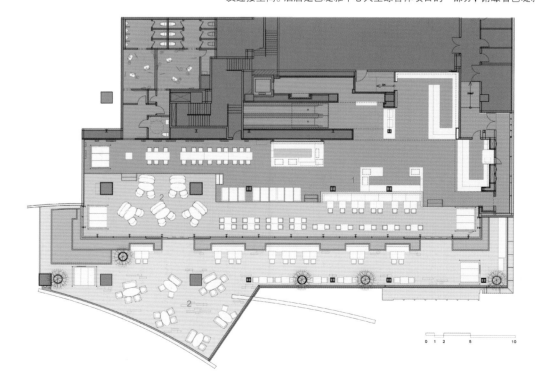

1. Open kitchen
2. Dining area

1. 开放式厨房
2. 就餐区

Edge Floor Plan 边缘餐厅平面图

"Edge" Restaurant

A restaurant serving international food with multiple large open kitchens is situated on the 13th floor facing the ocean view. Its main spatial organisation strategy is to open up the ocean view to the guests at its maximum to take advantage of the view from its prime location. The space is stretched linearly along the glass wall facing the sea with an almost 8 metre-high ceiling. The indoor seating area is organised into two tiers where the floor towards the back is higher up to ensure a good view over the front part. The outdoor terrace with its impressive panoramic view in the front lets the ocean vista flows uninterruptedly to the inside by arranging its floor plate a step lower.

Away from the busy street down below, the restaurant provides a calm, relaxing and comfortable atmosphere. An airy space with the use of natural materials and light colours allow the guests to set back and relax. The open kitchen area is accentuated with special material treatment as a focal point of the interior space.

The visual elements in the space are loose reminiscent of an underwater landscape – sea fan and translucent luminous ocean creatures. The interior surfaces are almost transformed from their original materiality into thin gorgonian membranes wrapping the space. Clusters of glowing organic-shape lamps suspended randomly in mid-air with varying sizes and colours scatter throughout the space. Hidden in the restaurant restroom, maneuvering through its interior space, one cannot resist thinking of the space in between the seabed fauna.

边缘餐厅

边缘餐厅位于酒店的14楼，面朝大海，专门供应国际美食，拥有多个大型开放式厨房。它的主要空间布局策略是最大限度地利用其优越的地理位置为宾客呈现海洋的美景。整个空间沿着朝向大海的玻璃墙展开，天花板足有8米高。室内就餐区分为两个层次，后排的地面略高，以便享有同等的海景。在露天平台能够将大海的景色一览无余，而略低的地面也为室内空间提供了良好的观景条件。

餐厅远离下方繁忙的街道，提供了宁静、放松、舒适的氛围。自然材料和浅色调所营造出的清新空间让宾客感到轻松愉快。开放式厨房所运用的独特材料是室内空间的焦点。

餐厅内的视觉元素引人联想起了水下景色——海扇和透明发光的海洋生物。室内的表面几乎全部贴上了柳珊瑚薄膜。一簇簇有机造型的吊灯在半空中悬垂下来，它们的大小和色彩都各不相同。无论是置身餐厅洗手间，还是漫步于室内空间，人们都无法忘记空间与海洋生物的联系。

As a focal point of the interior space, open kitchen is accentuated with special material treatment
作为室内空间的焦点,开放式厨房采用了特殊材料进行处理

Layerings of crystal, coloured sheer cloth, and wired-sculptured frame compose the lamp that gives extraordinary shadow to the interior space

吊灯由层叠的水晶彩色薄纱和铁丝框架结构组成，为室内空间带来了独特的光影效果

Natural material and light colour with natural sunlight gives a relaxing feel to the overall area

自然材料和浅色调与自然光照为空间营造出轻松的感觉

A cluster of glowing organic-shape light fixtures with crisscrossing-wooden membrane surface backdrop

一簇有机造型的吊灯与交叉的木条背景

View of the interior space with clusters of suspended lamps varying in sizes and colours　一簇簇大小、颜色不一的吊灯悬垂在半空中

Crisscrossing slat wall facing the casual dining space below marks an entrance to the fine dining　楼下休闲就餐区的交叉板条墙为楼上的高级餐厅打造了入口

Dining table with wine cellar background　餐桌，背景为酒架

Translucent volume of sheer fabric as a privacy screening yet connecting the spaces at the same time

半透明的薄纱形成了私密的屏风，同时也连接了空间

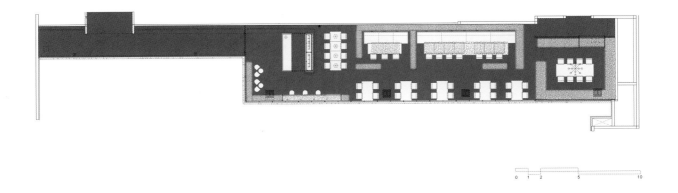

Floor Plan　平面图

"Flare" Fine Dining

An upscale luxury fine dining serving grill is located on a 14th floor. Privacy and exclusivity for the guests play an important role in the project. The design explores a mediating means of space demarcation between pockets of private dining area. By occupying an intermediary space between spaces with a translucent volume of sheer fabric, the effect results in an elegant, mystifying atmosphere, engaging and disengaging different spaces at the same time.

A deep colour palette gives an even further perplexing depth. A slight up-light to the volume of sheer fabric and the glowing light at the edge of the fabric accentuated the lighting in the space. The glitter of crystal lamps and chandeliers sparks and enriches the depth of the space.

闪耀高级餐厅

闪耀高级餐厅位于酒店的 15 楼，是一家豪华烤肉餐厅。餐厅设计的重点在于顾客的隐私和专属感。设计独创了一种划分私人就餐区的方式。设计师利用半透明的薄纱在不同区域之间创造了一个中介空间，形成了优雅、神秘的氛围，同时又分离了不同的空间。

深色色调赋予了空间复杂的深度。薄纱下方的微弱的灯光以及薄纱边缘的亮光为空间添加了独特的照明。水晶灯和吊灯的闪光丰富了空间的层次感。

Glittering elements within a deep interior colour

深色室内布景中的闪耀元素

Private dining space　包房

Assaggio Trattoria Italiana
阿萨吉欧意大利餐厅

Completion date: 2011
Location: Hong Kong, China
Designer: David Tsui, Suzanne Chong
Photos: Graham Uden
Area: 800m²

完成时间：2011年
项目地点：中国，香港
设计师：大卫·徐、苏珊娜·庄
图片版权：格雷汉姆·乌登
面积：800平方米

HASSELL was appointed by the Miramar Group to provide interior design consultancy for the Assaggio Trattoria Italiana Restaurant in Hong Kong, China. Assaggio Trattoria Italiana Restaurant is located on the fifth floor, inside the Hong Kong Arts Centre in Wanchai, adjacent to the Academy of Performing Art.

The restaurant has been well received by the customers and the general public since its opening in March 2011, and the completion of this project has created a great business relationship between HASSELL and the Miramar Group.

Sustainability
HASSELL's sustainable design approach is "Design with Nature". In order to introduce a green, natural, fresh and lively dining experience to the restaurant's dining guests, the design team used natural timber flooring, a recycled timber slab for the bar counter, old display cabinets as well as an herb garden outside.

米拉玛集团委托HASSELL设计公司为阿萨吉欧意大利餐厅进行室内设计。阿萨吉欧餐厅位于香港艺术中心的六楼，紧邻表演艺术学院。

餐厅自从2011年3月开张以来，深受顾客和公众好评。HASSELL设计公司也因此与米拉玛集团建立了长期的商业合作关系。

可持续设计
HASSELL设计公司的可持续宣言为"自然设计"。为了给餐厅顾客提供绿色、自然、新鲜、活跃的就餐体验，设计团队采用了实木地板、回收木板吧台和旧展示柜，并且在室外建造了一个香草园。

1. M&E service
2. PIZZA bar
3. AHU room
4. Light well
5. Kitchen
6. VIP room

1. 机械服务区
2. 比萨吧
3. 空调间
4. 光井
5. 厨房
6. 贵宾室

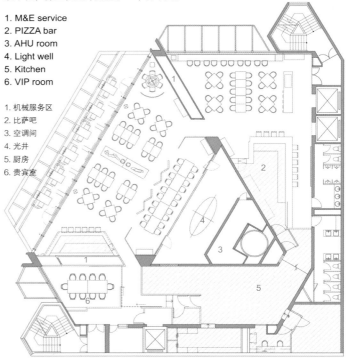

Floor Plan 平面图

Given its interesting location, the design concept of the restaurant is based on an artistic theatrical performance. Assaggio means "taste" and "little bites" in Italian language. The design concept aims to create a "dining stage" for the diners. The interior design is a deliberate attempt to make the venue look like a stage setting of a cosy Italian grocery store – and every paying customer in the restaurant is also seen as a part of the performance which tells a story about dining. The idea is to bring home a hearty Italian affair that transport diners to the friendly streets of Italy.

Deep wood hues make for a soothing dining scene and Italy is represented through vibrant art pieces fixed over rustic stone finishes. In addition, HASSELL proposed the idea of a pizza and pasta bar at the entrance of the restaurant to attract diners' attention upon their arrival and lead them to move to the Al Fresco area, which has a stunning view of the Victoria Harbour.

由于特殊的地理位置，餐厅的设计以戏剧表演为主题。在意大利语中，"阿萨吉欧"意为"品尝"和"浅尝"。餐厅设计旨在为就餐者打造一个"就餐舞台"。室内设计特别将餐厅打造成为一家舒适的意大利杂货店，而每一位前来就餐的顾客也成了表演的一部分。设计的理念是以正宗的意大利风情将就餐者带到意大利的热情街道上。

深色原木色调营造了舒缓的就餐场景，石板墙面上的艺术品则彰显了意大利风情。此外，HASSELL设计公司还在餐厅的入口添加了一个比萨意面吧，在一进门就吸引了顾客的眼球，引领他们进入俯瞰维多利亚港的露天就餐区。

Sugarcane

甘蔗餐厅

Completion date: 2010
Location: Miami, USA
Designer: Cetra Ruddy
Photos: Andrew Meade
Area: 446m²

完成时间：2010年
项目地点：美国，迈阿密
设计师：赛特拉·拉比
图片版权：安德鲁·米德
面积：446平方米

Springing from restaurateur SushiSamba's unique blend of Japanese, Brazilian, and Peruvian cuisine, music, and design, Sugarcane raw bar grill is a new tapas-style restaurant located in Miami's newly developing Design District. The restaurant/lounge features an 80m² indoor/outdoor bar and three distinct kitchens: robata, hot and raw bar. The spirited design is evocative of the Brasilian favela and features organic textures, salvaged ironwork, aged wood floors, reclaimed shutters, vintage "found objects", graffiti art and modern design, creating a sense of authenticity and age. Custom-designed banquettes, eclectic combinations of tables and chairs, hand-painted Moroccan tiles, antique mirrors and vintage fans create a comfortable yet festive atmosphere.

甘蔗生鲜烧烤餐厅融合了寿司桑巴餐饮品牌特有的日本、巴西和秘鲁美食以及音乐和设计元素，是一家位于迈阿密新设计区的新式西班牙小吃餐厅。餐厅拥有80平方米的室内外酒吧和三个独特的厨房：烤肉、热菜和生鲜吧。餐厅设计模仿了巴西的贫民区，以有机造型、回收铁制品、老化木地板、回收百叶窗、古董艺术品、涂鸦艺术和现代设计为特色，营造了真实而古老的感觉。定制卡座、精选的桌椅、手漆摩洛哥地砖、古董镜子和仿古风扇营造出舒适而欢乐的氛围。

The signature robata bar showcases the high flame cooking activity with bar seating for patrons to watch the chef at work. Both the indoor/outdoor and robata bars create a sense of drama and energy, allowing patrons to view each other and encouraging interaction.

Notable elements include the garden, robata bar and indoor/outdoor bar. The garden, hidden from the street by a hedge, continues the worldly experience through antique patio furniture, tropical plants, full height specimen trees and vines which cover the walls and window lattice ironwork. The indoor/outdoor was custom-designed with sliding glass doors that penetrate the centre of the bar, slipping behind a back-lit wall displaying the unique selection of sakes, tequilas and liquors.

招牌烤肉吧展示了大火烹饪技巧，顾客可以坐在吧台边观看大厨工作。室内外餐吧和烤肉吧的设计都充满了戏剧性和活力，顾客们可以进行交流和互动。

设计的重点元素在于花园、烤肉吧和室内外餐吧。花园通过树篱与街道隔开。露天家具、热带植物、园景树与覆盖墙壁和窗格的葡萄藤共同营造了愉快的体验。室内外餐吧采用了定制的玻璃拉门。拉门穿过餐吧中央，后方的背光墙上展示着精心挑选的日本清酒、龙舌兰和烧酒。

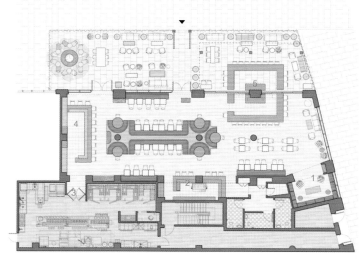

1. Lounge
2. Robarta bar
3. Kitchen
4. Raw bar
5. Indoor/outdoor bar

1. 酒吧间
2. 烤肉吧
3. 厨房
4. 生鲜吧
5. 室内外共享吧

Floor Plan 平面图

Sugarcane's variety of forms, styles and colours create a casual environment and communal experience that compliments the restaurant's initiative: for guests to kick back, eat well and stay awhile.

The favela design immediately immerses guests in an authentic and transformational experience of tropical culture and history. Multiple layers of multi-coloured paint create an aged patina which is echoed by the distressed furniture and finishes. The result is an atmosphere that is rich in character, embracing and timeless.

甘蔗餐厅以多样化的造型、风格和色彩打造了休闲的环境和社交体验，充分体现了餐厅的宗旨：让顾客吃好、喝好、休息好。

巴西贫民区式设计让顾客们宛如置身于真正的热带文化和历史。多层次的色彩叠加出老化的铜绿色，与老旧的家具和装饰相互呼应，最终形成了个性丰富、兼容并包的经典氛围。

Symphony's
交响乐餐厅

Completion date: 2009
Location: Amsterdam, The Netherlands
Designer: D/DOCK
Photos: Foppe Peter Schut
Area: 950m²

完成时间：2009年
项目地点：荷兰，阿姆斯特丹
设计师：D/DOCK设计公司
图片版权：福珀·皮特·舒特
面积：950平方米

With its multi-functional restaurant concept on Amsterdam's Southern Axis (Zuidas), Symphony's is meeting the huge demand from the many professionals in the area for a restaurant with an atmosphere and ambiance that holds its own with restaurants in the city centre. In cooperation with Vijverborgh – which developed the horeca concept – D/DOCK was responsible for the design and layout of Symphony's. A versatile restaurant that is a cross between a company restaurant and an establishment that is open to the public.

交响乐餐厅以多功能餐厅为概念，位于阿姆斯特丹的南侧轴线之上，满足了该地区职业人士对餐厅的巨大需求。餐厅的环境和氛围充分体现了市中心的气息。D/Dock 设计公司与 Vijverborgh（负责餐具设计）合作，负责餐厅的设计和布局。交响乐餐厅兼具企业餐厅和豪华餐厅的特点，向公众提供各色美食。

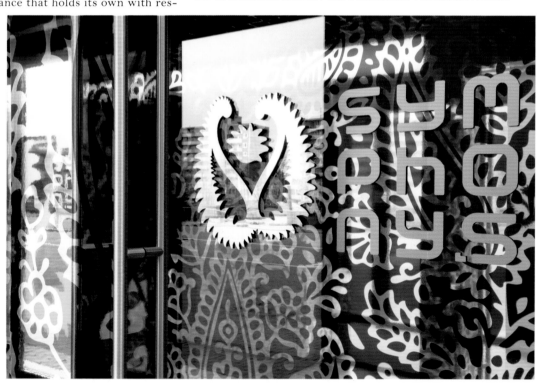

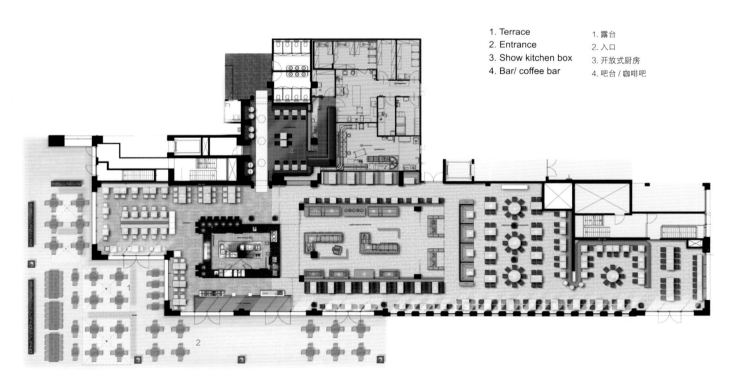

1. Terrace
2. Entrance
3. Show kitchen box
4. Bar/ coffee bar

1. 露台
2. 入口
3. 开放式厨房
4. 吧台 / 咖啡吧

Ground Floor Plan   一层平面图

In the heart of the restaurant, there is a show kitchen in which the chefs cook in full view of the guests, creating a feeling of authenticity, freshness and dependability.

The multi-functionality of Symphony's, which is embedded in the layout and design of the concept, makes it an attractive port of call throughout the day for company employees and people from the surrounding offices. The same space where coffee and breakfast are served early in the mornings transforms into a popular cocktail bar in the evenings.

在餐厅中央的开放式厨房里,厨师向顾客展示着烹饪的全过程,营造出真实、新鲜、可信的感觉。

交响乐餐厅的多功能性体现在布局和设计理念上,吸引了大量的周边公司员工前来就餐。餐厅的吧台在早晨供应早餐和咖啡,晚上则化身为鸡尾酒吧。

Symphony's has been turned into a comfortable and fashionable restaurant with a characteristic layout based on the "Techno India" theme. The unusual use of colour and the recurring paisley pattern set the restaurant from its options for use.

餐厅设计以"印度电子乐"为主题,营造了舒适、时尚的空间。独特的色彩和不断重复的涡纹图案奠定了餐厅的基调。

De Gusto

好味道品鉴餐厅

Completion date: 2009
Location: Ravenna, Italy
Designer: Arch. Camilla Lapucci & Lapo Bianchi Luci
Photos: Pietro Savorelli
Area: 300m²

完成时间：2009年
项目地点：意大利，拉文纳
设计师：卡米拉·拉普希与拉波·碧昂琪·路西
图片版权：彼得罗·萨沃雷利
面积：300平方米

De Gusto can be defined an "Italian pasta University", it is a tasting hall located inside Surgital industrial complex in Lavezzola (RA), leading company in deep frozen pasta exportation.

The hall is a culinary laboratory and it represents the heart of the "Research & Development" department. It is really important because it is the kitchen where Surgital can connect with its customer and promote its work.

The opening party was directed by the international chef Gianfranco Vissani and many public officers were invited to the event. A special menu was created and cooked inside De Gusto kitchen, pasta was obviously the main element. The menu was richly completed with "Divine Creazioni", the excellent fine line.

The idea of De Gusto rises up from the necessity of creating a representative room as a memorable end of the visiting path of the company. Everyone can observe excellent chefs cooking.

"Every week we host restaurant owners and usually we offer tasting samples of our product at the end of the visit" says Elena Bacchini – Surgital Marketing Director – "we usually contact external famous chef to create culinary experimental events".

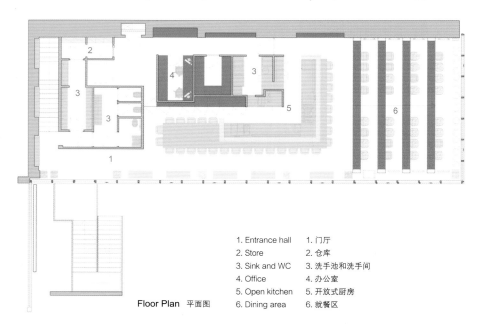

Floor Plan 平面图

1. Entrance hall　1. 门厅
2. Store　2. 仓库
3. Sink and WC　3. 洗手池和洗手间
4. Office　4. 办公室
5. Open kitchen　5. 开放式厨房
6. Dining area　6. 就餐区

好味道品鉴餐厅堪称"意大利面大学",它位于拉文纳的苏吉塔公司——一家顶尖的速冻意面出口公司的工业大楼内。

餐厅属于一间烹饪实验室,是研发部门的中心。餐厅的重要性体现在它是苏吉塔公司与顾客交流的平台,也是宣传其产品的主要方式之一。

国际名厨詹弗兰科·维萨尼主持了餐厅的开幕派对,到场的还有许多政府官员。好味道厨房特别制作了精美的菜肴,而意面当然是主角。菜单内特别添加了"神圣创作"———一种出色的细面条。

好味道餐厅的设计起源于公司需要为参观路线打造一个令人难忘的终点。人们可以在餐厅内欣赏大厨的精湛厨艺。

"我们每周都会邀请外面的餐厅业主前来参观,并且以品鉴我们的产品作为参观的终点",苏吉塔公司市场总监称,"我们经常联系优秀的厨师前来主持烹饪体验活动。"

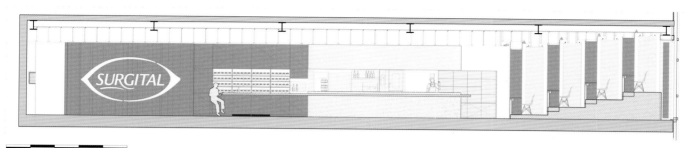

Section 剖面图

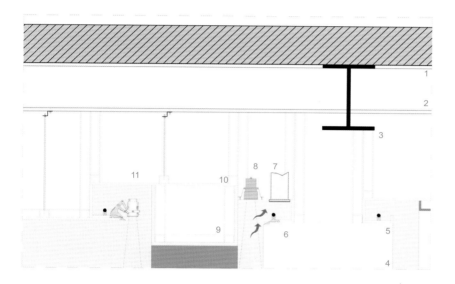

1. Roof structure
2. Steel structure
3. Cross beam
4. Plasterboard panel for communication
5. Light linear diffuser hidden into plasterboard ceiling
6. Light linear diffuser on railed metallic profile
7. Linear air vent
8. Recessed spotlight
9. Wood portal
10. Plasterboard structure
11. Railed spotlight

1. 屋顶结构
2. 钢结构
3. 横梁
4. 通信石膏板
5. 石膏天花板后面的线型漫射灯
6. 金属轨道上的线型漫射灯
7. 线型通气孔
8. 嵌入式射灯
9. 木门
10. 石膏板结构
11. 轨道射灯

Ceiling detail 天花节点图

A technologic heart of professional devices is the real heart of the kitchen, covered with calacatta marble, the one traditionally used as table to prepare hand-made pasta. Next to it there is a 15m long table with thirty stools to watch the preparation from the first line.

Main characters of De Gusto are materials and textures. Grey concrete floor, porous brushed oak wood furniture, calacatta marble veins and Japanese barely wall paper create a domestic and familiar atmosphere that exalt pure, essential volumes.

餐厅的技术核心在于厨房的中心。厨房内的台板由白色大理石——制作手工意面的传统桌面——制成。15米长的桌子紧紧围绕着台板，四周摆放着30张高脚凳，人们可以坐在上面观看意面准备过程。

好味道餐厅的特色在于建筑材料和纹理。灰色水泥地面、拉丝橡木家具、卡拉考特大理石纹理与日式素面墙纸共同营造出家居氛围，整个空间显得纯粹自然。

Nature and Technology are strictly connected: a vertical garden goes with visitors along entrance stairs and inside all fine senses are stimulated.

The interior design is functional, rigorous but minimal and simple. Each one can easily individuate all functions and enjoy them naturally. The open space is divided into a technologic interactive kitchen and in a sort of big lecture hall. Couples of long wooden tables, where about eight people can take a sit, are placed each one on a step of a huge stair in front of the kitchen. Visitors can observe excellent chefs cooking and then taste their creation prepared with Surgital product.

De Gusto confirms the high quality of "Made in Italy", as Surgital does it constantly exporting its products all over the world. The great result is due to the ability of architects C+L Studio to mix traditional culinary elements, proposed in an upgraded contemporary layout, and technologic devices, audio and video system.

自然与技术紧密相连：楼梯入口的垂直花园为参观者带来了良好的感觉。

室内设计严谨实用，且简洁明快。任何人都能区分各个功能且熟练运用。开放空间被划分为技术互动厨房和大型演讲厅。一些能容纳8人的长木桌摆放在厨房前方的大台阶上。参观者可以边观看大厨的精湛厨艺边享用美味的苏吉塔产品。

好味道餐厅旨在突出"意大利制造"的优良品质，正如苏吉塔公司一直致力于将它的产品出口到全世界一样。C+L工作室将传统烹饪元素与现代布局相结合，并且融入了技术设备、视频和音频系统等。

161

Karls Kitchen
卡尔斯小厨

Completion date: 2010
Location: Stuttgart, Germany
Client: E. Breuninger GmbH & Co.
Designer: Concrete Architectural Associates
Project team: Rob Wagemans
      Melanie Knüwer
      Charlotte Key
      Erik van Dillen
      Sofie Ruytenberg (graphic design)
Photos: Ewout Huibers
Area: 1,000m²
Seats: 400

完成时间：2010年
项目地点：德国，斯图加特
委托人：伯宁格公司
设计师：混凝土建筑事务所
项目团队：罗伯·维格曼斯
    梅兰妮·努维尔
    夏洛特·吉
    埃里克·范迪伦
    索菲·陆登伯格（平面设计）
图片版权：艾沃特·修伯斯
面积：1,000平方米
座位：400

Karls Kitchen is the new restaurant of luxury department store Breuninger in Stuttgart.

The kitchen is the free-flow restaurant's central element. The entrance to the restaurant is characterised by large glass fridges with fresh products that also function as a display window.

Materials:
Floor: The entire restaurant is fitted with smooth, oak floors, with a different finish for each seating area (white, grey and black).
· Kitchen: Terrazzo tiles (colour custom-made), 60x60cm
· Toilets: Floor tiles Mosa, beige and brown, 15x60cm
Ceiling: The entire restaurant is fitted with a white acoustic plasterboard ceiling, except the more intimate areas have a dark ceiling. The lounge is fitted with a dark wood vertical panel ceiling; the kitchens have a black metal ceiling and the private kitchen has its ceiling painted black.
Wall: Custom-made wallpaper, Concrete's design produced by Berlintapete.
Interior: Custom-made furnishings, such as the "Stammtisch", the children's corner, the sofas and the tables, were designed by Concrete and produced by Van Bergh, interior builder.
Lighting: The green hanging lamps in the modern section were specially designed by Concrete and produced by Baulmann.
In the lounge area, the hanging lamp by Raimond van Moooi and the Base standard lamp from Tom Dixon create the intimate atmosphere.
The porcelain lamps above the "Stammtisch" come from a Dutch ceramics studio.
In the kitchen section and the walking areas, spotlights with close lighting bundles provide a theatrical effect.
Furniture: There is a choice of various chairs for each guest. Nearly all the tables and chairs have something classical with a modern interpretation. The chair and the InOut bar stool from Gervasoni have a very simple form but are made of aluminium. The Morph van Zeitraum chair in the modern seating area is made of walnut.
The tables and AVL benches in the traditional section are from Moooi.
The Ray black leather lounge chairs and the DLM metal tables are from Hay. The outdoor furniture from the Tio series is from Massproduction.

卡尔斯小厨是斯图加特伯宁格奢侈百货店内的一家新餐厅。

厨房是餐厅的核心元素。餐厅的入口—巨大的玻璃桥为特色,玻璃桥内部展示着新鲜的食品,也相当于橱窗。

材料:
地面:整个餐厅都铺设着光滑的橡木地板,每个座位区的色彩各不相同(如白色、灰色和黑色)。
• 厨房:水磨石地砖(定制色),60x60cm
• 洗手间:莫萨地砖,米黄和棕色,15x60cm
天花板:餐厅大部分空间都配置着白色隔音石膏板吊顶,部分私密区域为黑色吊顶。例如,休息区配有黑色木制垂直面板吊顶,厨房采用了黑色金属吊顶,而小厨房的天花板则被涂成了黑色。
墙壁:定制墙纸(混凝土建筑事务所设计,柏林塔比特制作)。
室内:定制家具,如常客桌、儿童角、沙发和桌子(混凝土建筑事务所设计,范伯格制作)。
照明:现代区的绿色吊灯由混凝土建筑事务所设计,鲍尔曼制作。
休息区的吊灯由雷蒙德·范莫伊设计,标准底灯由汤姆·迪克森设计。
常客桌上方的陶瓷灯来自一个荷兰陶瓷工作室。
厨房和步行区的聚光灯所射出的光束营造了剧院的效果。
家具:顾客有多种座椅可以选择。几乎所有桌椅都融合了古典和现代。基尔瓦索尼制作的座椅和吧台长凳造型简单,由铝材制成。
现代区的莫夫·范吉特劳姆座椅采用了胡桃木。
传统区的餐桌和长椅来自莫伊。
黑色皮革座椅和DLM金属桌来自黑尔设计。露天家具则属于批量生产的迪欧系列。

THE KITCHENS

The guest's initial experience is on coming in, walking through the kitchens, barely separated from the preparation of fresh products by a large glass wall.

To the left you can see exclusive sandwiches, fresh salads and homemade desserts being prepared. On the right you can take a peek into the warm kitchens, where traditional local classics, Asian and European specialities are prepared. The dishes, which change daily, are presented on a row of screens above the two kitchen counters.

The materials used reference the classic kitchen of grandma's time: white tiles, black natural stone in combination with a terrazzo floor. The modern design in combination with new materials such as glass and stainless steel create a modern design with a link to the classic kitchen.

厨房

顾客对餐厅的第一印象就是厨房。各个不同的厨房通过大型玻璃墙内的新鲜出炉的美食相互隔开。

左侧的厨房制作特制三明治、新鲜沙拉和手工甜点。右侧的热厨房烹制当地传统美食以及亚洲和欧洲小吃。厨房柜台上方的一排屏幕上展示着每日经典菜肴。厨房装饰材料选择与传统私家小厨类似:白瓷砖、黑色天然石材和水磨石地面。玻璃、不锈钢等新材料的运用将现代设计与传统厨房结合起来。

THE BAR

Opposite the kitchens is the bar, a free-standing element between the free-flow restaurant and the seating area. The bar's function is two-fold: on the kitchen side, cakes and pastries are made in-store; fresh juices and beverages are offered.

The bar adjacent to the lounge area tempts one to sit at the bar and enjoy a nice glass of wine, a glass of local beer or one of the coffee specialities.

The materials used for the bar create a connection between the restaurants and the kitchens. The bar is made of wooden tiles in the same dimensions as the kitchen tiles, on top of which is a black natural stone surface, creating a warm appearance and being functional at the same time.

酒吧

酒吧正对厨房，是独立于餐厅和休息区的独立元素。酒吧的功能具有两面性：厨房为酒吧准备好蛋糕和甜点；酒吧负责提供鲜果汁和酒水。

紧邻休息区的酒吧吸引着人们前来品尝红酒、啤酒或特制咖啡。

酒吧的装饰材料连接了餐厅与厨房。吧台由与厨房瓷砖相同尺寸的木砖支撑，顶部铺有黑色天然石材表面，既显得温馨，又十分实用。

1. Warm kitchen (80m²)
2. Cold kitchen (50m²)
3. Bar (55m²)
4. Traditional restaurant (100m²)
5. Contemporary restaurant (195m²)
6. Private kitchen (43m²)
7. Dishwashing area (32m²)

1. 热厨房（80m²）
2. 冷厨房（50 m²）
3. 酒吧（55 m²）
4. 传统餐厅（100 m²）
5. 现代餐厅（195 m²）
6. 私人厨房（43 m²）
7. 洗碗区（32 m²）

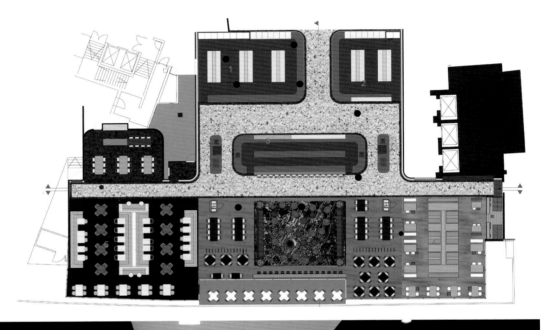

Floor Plan　平面图

The Lounge

The Lounge area in the middle of the space is directly adjacent to the bar. The lounge has a relaxed atmosphere with low lounge chairs and an extra long black leather sofa. In the middle a hanging fireplace provides an intimate feel. The lounge-character is reinforced by a lowered black vertical strip ceiling and a dark-grey wooden floor.

The lounge is linked directly to the terrace by means of the vertical panel ceiling which runs from inside to outside. The outside panels also serve as an awning.

The terrace's wooden floor is continued inside, forming a long table along the façade, allowing one to enjoy the view of Stuttgart from indoors as well.

休息区

中央休息区紧邻酒吧。低矮的休闲椅与超长的黑色皮革沙发营造出轻松的氛围。中间的壁炉令人感到亲切温暖。低矮的黑色竖条天花板和黑灰色地板进一步突出了休闲的氛围。休息区通过贯穿内外的天花板与露台直接相连。外面的天花板还起到了雨篷的作用。

露台的木地板与室内相连，沿着外墙形成了一张长桌，让人们在室内也能欣赏斯图加特的美景。

THE RESTAURANT

The restaurant is divided into 3 different seating areas, each with its own identity and character. First of all there is the traditional seating area, which refers to Breuninger's history and the traditions within the department store, as well as in Stuttgart. The modern seating area represents the Stuttgart of today, and the Lounge is the refuge and, at the same time, the connection between the modern and the traditional seating areas.

An overarching element connecting the spaces to one another is one long wall with graphic wallpaper. Where walls are interrupted by glass panels or doors, the glass is decorated with the same print, but transparent, thereby maintaining the overall image.

The Traditional Restaurant

The traditional part is a modern interpretation of an original "Bierstube". The central element here is the "Stammtisch" (table for the regulars). This is one big table from which pieces have been cut for the sofas. The solid-oak top contrasts with the white legs, which are made of Corian. In total the table provides seating space for 34 people. In the middle of the table there is, just like in the modern section, a high cabinet up to the ceiling.

This cabinet is filled with coloured red products such as kitchenware and local souvenirs. The red recurs in a number of chairs, which provide a hint of colour between the white furniture.

Above each table hang two lamps. These resemble a classic embroidered lampshade from grandma's time. What is unique about the lamps is that they are made of white porcelain with a red iron cable. It thereby ties in with the concept of an old-fashioned product with a contemporary twist.

The white bare oak floors of the traditional section reinforce the light atmosphere. The children's play corner is right next to the traditional seating area. It is a kind of open box made of the same white wood as the floor. Red cushions, beanbags and stickers with red animals on the glass side walls create a playful touch.

There is a dedicated parking place for pushchairs with a parking meter.

餐厅

餐厅分为三个座位区，它们各有特色。传统座位区参照了伯宁格百货乃至斯图加特的历史和传统。现代区代表着斯图加特的现在。而休息区则相当于一个避难所，同时也连接了传统和现代两个区域。

贴有印花墙纸的长墙将这些空间连接起来。在墙面上的玻璃板或门板上，采用了同样图案的透明印花，保持了整体造型。

传统餐厅

传统餐厅是现代版的德国啤酒屋。中心元素是一张"常客桌"。这张巨大的长桌经过切割，融入了沙发的设计。硬实的橡木桌面与白色人造石桌腿形成了对比。这张桌子可以为34个人提供座位。桌子的中间采用了与现代区相同的设计，设置了一个直达天花板的橱柜。

橱柜内摆满了红色系的商品，如餐具、当地纪念品等。一些座椅也采用了红色装饰，在雪白的家具中起到了点缀作用。

每张餐桌上方都有两盏吊灯。吊灯采用了古典家居式的刺绣灯罩。吊灯的独特之处在于它们是由白色陶瓷和红色电缆组成的，紧贴老式风格与现代品味相结合的主题。

白色橡木地板突出了明亮的氛围。儿童游戏角紧邻传统区。儿童角类似一个白木制成的盒子。红色靠垫、豆子袋和动物贴纸营造了欢快的感觉。

此外，餐厅还设有婴儿车的专属停车位，车位采用停车计时器收费。

The Modern Restaurant

The modern seating area is defined by contemporary materials, such as Corian, light-coloured leather, chrome in combination with dark wood.

The central element is a furniture piece in the middle of the room. It comprises long sofas with light leather upholstery connected by a high cabinet. In this cabinet there are cooking utensils in silver and chrome which reflect towards the modern character.

The dark wooden floor creates a subtle contrast with the light furniture.

The custom-made hanging lamps, with the different dimensions and shades of green, provide a playful touch.

Private Kitchen

Alongside the three seating areas there is the restaurant's private kitchen, which can be closed off by means of a sliding glass door. The room has its own free-standing cooker and can be used for cooking workshops, parties, presentations or business meetings. The materials, such as the white tiles and the kitchenware along the walls, partly come from the kitchen. The dark wooden floor and a black ceiling create an intimate, informal atmosphere.

现代餐厅

现代座位区采用了具有现代感的材料，如人造大理石、浅色皮革、铬合金与黑木的结合等。

中心元素为房间中央的组合家具。它结合了浅色皮革长沙发和高橱柜。橱柜上摆放着银和铬合金的厨房用具，反映了现代风格。

黑木地板与浅色家具形成了微妙的对比。

定制吊灯大小不一，采用了不同深浅的绿色色调，趣味十足。

私人厨房

三个座位区旁边是餐厅的私人厨房，厨房通过玻璃拉门进行封闭。房间内设有独立的炊具，供烹饪教室、排队、展示或商务会议使用。私人厨房的装饰材料有一部分来自于开放式厨房，如白瓷砖和沿墙的厨具。黑色木地板和黑色天花板营造出私密、放松的氛围。

The River Café

河畔咖啡馆

Completion date: 2008
Location: London, UK
Designer: Stuart Forbes Associates / Stuart Forbes and Richard Rogers
Photos: Grant Smith
Area: 468m²

完成时间：2008年
项目地点：英国，伦敦
设计：斯图亚特·福布斯事务所/斯图亚特·福布斯与理查德·罗杰斯
图片版权：格兰特·史密斯
面积：468平方米

The River Café, one of London's most celebrated restaurants, re-opened its doors on 6 October 2008 after a £2 million re-design. Following a fire in April 2008, co-owners Ruth Rogers and Rose Gray commissioned Stuart Forbes Associates to re-design the space with a new open kitchen, private dining room and cocktail bar.

Since the re-launch, the restaurant has enjoyed the acclaim of the world's food critiques and architectural press and continues to be one of the most influential food establishments.

作为伦敦最负盛名的餐馆之一，河畔咖啡馆在耗资 200 万英镑进行重新设计之后，于 2008 年 10 月 6 日重装上阵。由于咖啡馆在 2008 年 4 月遭遇了火灾，合伙人卢斯·罗杰斯和罗丝·格雷委托斯图亚特·福布斯事务所对其进行重新设计，以开放式厨房、包房和鸡尾酒吧为重点。

重新开张的餐馆获得了世界美食评论界和建筑评论界的一致好评，被认为是最具影响力的食品机构之一。

1. Main entrance
2. Bar
3. Coat room
4. Wine storage
5. Restaurant
6. Terrace
7. Garden
8. Desserts
9. Cold foods
10. Kitchen
11. Wash up
12. Private dining
13. Cheese store
14. Staff changing
15. Deliveries entrance

1. 主入口
2. 吧台
3. 衣帽间
4. 酒窖
5. 餐厅
6. 露台
7. 花园
8. 甜点区
9. 冷餐区
10. 厨房
11. 洗碗间
12. 包房
13. 芝士商店
14. 员工更衣室
15. 送货入口

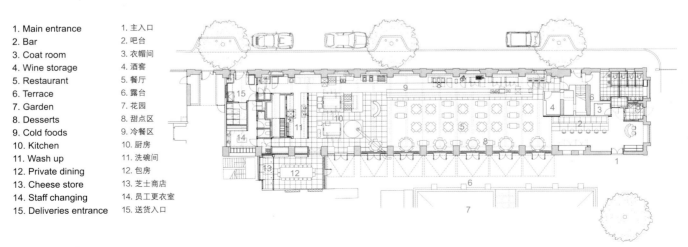

Floor Plan 平面图

Significant additions include a kitchen table within the working kitchen zone and a glass enclosed cheese room based on a classic Italian design. The kitchen is open to public view, enabling diners to observe the chefs in action.

Another notable addition is a new wood burning oven dominating the dining room and kitchen. Manufactured by the Italian firm of Valoriani, this type of oven, made in Reggello in the hills of Tuscany 25km southeast of Florence, is key to the preparation of the The River Café's many dishes.

The reception desk and "pass" have been constructed in vivid yellow "Corian" and dramatically top-lit. This new area is located next to The River Café bar which has been adapted and enhanced with a new back-lit bottle display. Behind the scenes, the bathrooms have been re-planned and reconstructed with fluorescent primary colours and significantly increased space.

设计在厨房工作区内添加了一张台板，并且新增了一个经典意大利风格的玻璃芝士屋。厨房面向公众开放，人们可以看到厨师的活动。

新增的木柴烤炉是餐厅和厨房中最显眼的元素。烤炉由意大利瓦罗里阿尼公司生产，在佛罗伦萨东南25千米的托斯卡纳山区的雷杰洛制成，是制作河畔咖啡馆经典菜肴的必需品。

餐厅的前台接待处由活泼生动的黄色人造大理石制成，并采用了引人注目的顶部照明。旁边的吧台经过改造，采用全新的背光照明进行酒品展示。后方的洗手间经过重建，以荧光色为基调，并且扩大了内部空间。

Section A 剖面图 A

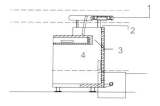

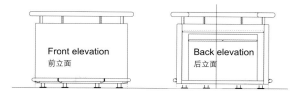

Front elevation 前立面

Back elevation 后立面

Oven Design
烤炉设计

Section A
1. All white
2. Yellow
3. Open access to back of equipment
4. Power sockets

剖面图 A
1. 全白
2. 黄色
3. 通往后方设施的开口
4. 电源插座

Plan 1 设计 1

Plan 2 设计 2

Plan 3 设计 3

Plan 1
1. Brackets and plates sizes
2. Curved to stop here at 80mm from front edge. Then to become straight to achieve a clear front edge

设计 1
1. 支架和板面尺寸
2. 弧度距前缘 80mm 处结束，变为直线，形成清晰的前缘

Plan 2
1. Φ75mm cut out to allow of cables connection. Provide edge detail

设计 2
1. Φ75mm 开口，安装电缆接头。提供边缘细部设计

Plan 3
1. Pull out tray for keyboard
2. 515mm equal on both sides

设计 3
1. 抽出托盘，放置键盘
2. 两侧均为 515mm

Detail section 细部剖面图

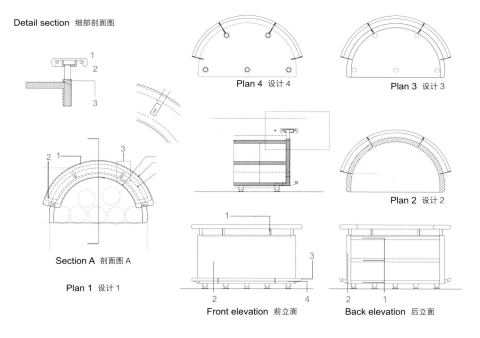

Plan 4 设计 4

Plan 3 设计 3

Plan 2 设计 2

Section A 剖面图 A

Plan 1 设计 1

Front elevation 前立面

Back elevation 后立面

Detail section
1. Curved Top to allow for lighting feature Hout Tek to liason with lighting consultant for appropriate spacing connection details and transformers location on the desk
2. S/S Posts 50x8mm to connect Curved Top with Bottom Foot
3. 25mm Upstand with 20mm rounded junction to counter top

细部剖面图
1. 弧形顶，以便安装特色照明 Hout Tek 与照明顾问共同确定合适的间距细节和桌面转换器的位置
2. 50x8mm S/S 柱，连接弧形顶和底脚
3. 25mm 竖柱，配有 20mm 圆角接点连接台面

Section A
1. Lighting feature
2. Curved top area
3. Fixing Plate between Vertical S/S Post and underside of Curved Top

剖面图 A
1. 特色照明
2. 弧形顶区域
3. 固定板 S/S 柱和弧形顶底面之间的固定板

Front elevation
1. S/S Posts 50x8mm to connect Curved Top with Bottom Foot
2. "Corian" or similar approved
   Colour: Blue to be confirmed by sample provided by Hout Tek
3. Φ50mm St.Steel 304 Tubular Footrail on St.Steel Bracket
4. Φ50mm S/S Posts Legs on Φ100mm Stability Plate. Fixed to desk underside and connected to vertical posts as required

前立面
1. 50x8mm S/S 柱，连接弧形顶和底脚
2. "可丽耐" 或类似材料
   色彩：Hout Tek 提供的蓝色样板
3. Φ50mm St. 钢 304 管状横档，安装在 St. 钢支架上
4. Φ100mm 稳定板上的 Φ50mm S/S 柱腿。固定在桌底面，与立柱连接

Back elevation
1. "Corian" or similar approved
2. Colour: Blue to be confirmed by sample provided by Hout Tek

后立面
1. "可丽耐" 或类似材料
2. 色彩：Hout Tek 提供的蓝色样板

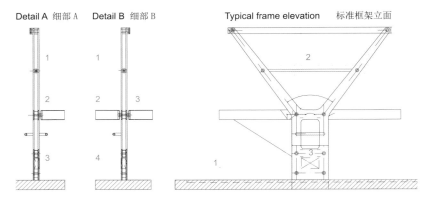

Detail A
1. Optional additional pan shelf
2. Shelf cavity with recessed lighting
3. Flush mounted S/S double electrical socket facing inward towards the pre surface

细部 A
1. 可选附加锅架
2. 架子内部，配有嵌入照明
3. 嵌装 S/S 双插座，朝里

Detail B
1. Optional additional pan shelf
2. Shelf cavity with reccssed lighting
3. Shelf cavity with reccssed lighting
4. Flush mounted S/S double electrical socket facing inward towards the pre surface

细部 B
1. 可选附加锅架
2. 架子内部，配有嵌入照明
3. 架子内部，配有嵌入照明
4. 嵌装 S/S 双插座，朝里

Typical frame elevation
1. Work surface
2. 20mm solid bar
3. Cut out for 2x electrical plug socket

标准框架立面
1. 工作台
2. 20mm 实心棒
3. 双插头插座切口

General note
All fabricated steel to be S/S Grade 316
Brushed finish to match worktops
All through bolts to be countersunk hexagon socket machine screws
Cut plate steel to bolt onto 60x30mm box section frame fixed to base table
All steel sizes shown are indicative and must be verified by the contractor

附注
所有结构钢均为 S/S 级 316
工作台需进行刷面处理
所有双头螺栓均为埋头六角机械螺丝
将钢板切割成 60×30mm 盒形断面，固定在底座上
所有钢材的尺寸均为指导性的，需经过承包商复核

There are dining areas of different styles in the restaurants, which add the flexibility and diversity of the interior and provide more choice for diners.

餐厅内的就餐区风格各异，打造了灵活多样室内设计，并且为就餐者提供了更多选择。

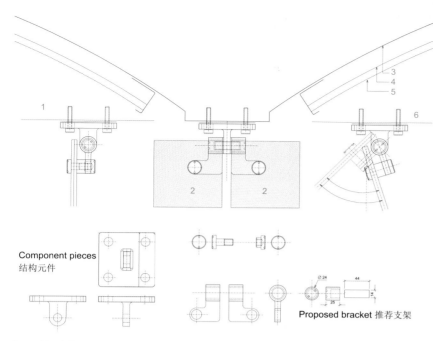

Component pieces 结构元件

Proposed bracket 推荐支架

Assembly of glass panel to bracket and structural soffit 支架和结构拱面的玻璃板装配

Note
The dimensions shown on the original square plate and eye are taken from the existing brackets in the River Café. The original glass specification is 10mm toughened monolithic sheets. In the event that this specification changes to a laminated panel, the unit weight must be taken into account when proof checking the bolt fixings and bracket back to the structural soffit.
The bolt fixings to the structural soffit are drilled and tapped into the flange of the steel beams of the original building fabric.
The fixing method shown is for illustrative purposes as "design intent" and must be verified by a structural engineer.
All component pieces are fabricated in stainless steel.

附注
原始方形板上所显示的尺寸数据来自咖啡厅的原有支架。原始玻璃规格为10mm加强整体板。设计将其改成了夹层板，因此在验证结构拱面的螺栓固定和支架时必须考虑到单位重量。
结构拱面的螺栓固定采用在原有建筑钢梁上钻孔的方式。
所显示的固定方式为"设计参考"，需经过结构工程师复核。
所有元件均以不锈钢组装。

1. Panel shown vertical
2. Laminated glass panel
3. Structural soffit
4. Acoustic quilt
5. S/S weld mesh ceiling panel
6. Range of panel movement

1. 玻璃板的垂直展示
2. 夹层玻璃板
3. 结构拱面
4. 隔音层
5. S/S焊接网格天花板
6. 隔板运动的幅度

Bar Brasserie Restaurant Fitch & Shui

费奇&舒伊酒吧餐厅

Completion date: 2011
Location: Amsterdam, The Netherlands
Designer: D/DOCK
Photos: Ronald Nieuwendijk
Area: 1,400m²

完成时间：2011年
项目地点：荷兰，阿姆斯特丹
设计师：D/DOCK设计公司
图片版权：罗纳德·纽恩迪治克
面积：1,400平方米

Located in the Amsterdam WTC building, the Fitch & Shui restaurant is the eye catcher from the front side of the building. The big glass façade gives a passers-by free view inside on the long bar and the cozy seating at the window. The elongated space asked for a big gesture like the bar to define the space. The setup of the bar also symbolises its use: starting with breakfast, fresh juice, sandwiches for lunch, cocktails at the bar and a big "wine painting" at the end.

费奇&舒伊酒吧餐厅位于阿姆斯特丹世贸中心大楼，十分引人注目。大面积的玻璃外墙让路人可以透过窗户任意观看内部的吧台和舒适的座椅。细长的空间需要长吧台来突出重点。酒吧的布置反映了它的功能：从早餐开始，到午餐的鲜果汁、三明治，再到吧台上的鸡尾酒和端墙上的巨幅"红酒壁画"。

Together with the hanging meat and other fresh products on and behind the bar, the bar is made to be a "food theatre", where honest products are sold by experience, completely in line with the Fitch & Shui concept.

酒吧内悬挂着肉制品，吧台上面和后方也摆放着其他新鲜的食品，看起来就像一家"食品剧院"。酒吧以丰富的经验贩售保真的商品，与费奇＆舒伊的主题完全一致。

Section 剖面图

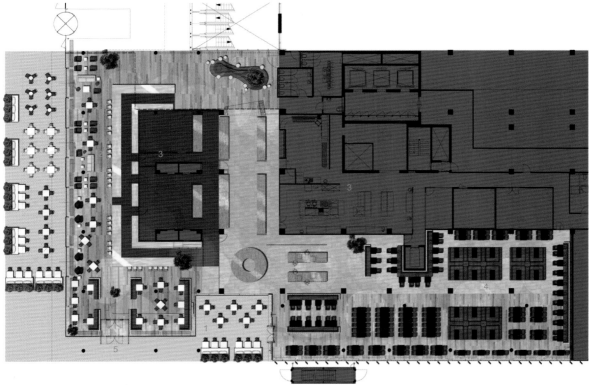

1. Terrace
2. Bar
3. Kitchen
4. Dining area
5. Entrance

1. 露台
2. 吧台
3. 厨房
4. 就餐区
5. 入口

Floor Plan 平面图

The self-service restaurant has a free-flow setup and has seating for 300 people. The interior provides different seating typologies: high, low, half-high, bench, chair and stool. These variations in seating possibilities divide the space in areas and the different modules give protection in the big open space, in a simple cost effective way.

Along the windows a variety of furniture like big couches, fauteuil's, stools, big lamps and several mood and styling elements are set up in a playful way. The space as a whole breaths a warm cosmopolitan ambiance with an old-skool twist.

自助餐厅采用开放式布局，可容纳300人同时就餐。室内设有多种座位类型：高椅、矮椅、半高椅、长椅、普通椅和高脚凳。这些不同类型的座椅区分了不同的空间，而不同的模块则保护了大型开放空间。这一设计简单有效，又经济实惠。

靠近窗户的位置摆放了一系列不同的家具：长沙发、扶手椅、高脚凳、大吊灯和不同风格的元素。整个空间散发出温馨而古典的氛围。

The walls are cover with a graphic illustration of herbs, specially designed for Fitch & Shui, by D/DOCK. In leather-upholstered couches, black or wooden tables and spherical lamps are placed throughout the space, enhancing the feeling or the different protected modules. All these elements give the self-service restaurant a simplistic modern classic appearance, with an industrial touch.

墙壁上的平面图案有 D/DOCK 专为费奇 & 舒伊餐厅设计。皮革软沙发内部摆放着黑色或木制餐桌和球形灯，增强了空间的模块感。这些元素共同为自助餐厅营造了简洁现代的外观，带有独特的工业化风情。

FRESH CUTT
新鲜出炉烧烤店

Completion date: 2009
Location: Sherman Oaks, USA
Designer: AkarStudios
Photos: Randall Michelson
Area: 223m²

完成时间：2009年
项目地点：美国，谢尔曼橡树地区
设计：阿卡工作室
图片版权：兰德尔·迈克逊
面积：233平方米

Featuring a dramatically lit open kitchen and serving counter, the quick-service Mediterranean concept is a result of a major transformation of a space that previously housed a salad-bar concept. The former venue was completely altered to create a sense of freshness that would appeal to an aspirational audience. With an array of contemporary eco-friendly materials and finishes, the design has been kept clean and simple, with the main counter becoming the focal point. AkarStudios created a branded identity and graphics programme for the launching of this new concept.

这家地中海快餐店以明亮的开放式厨房和服务台为特色，由一家沙拉吧改造而成。餐厅经过全面改造，营造了一种能够吸引挑剔食客的新鲜感。现代环保材料和装饰的使用让设计显得简洁干净，突出了主服务台的中心位置。阿卡工作室为餐厅设计了品牌形象和对应的图案标识。

1. Kitchen　　　　　1. 厨房
2. Exhibition kitchen　2. 展示厨房
3. Prep counter　　　3. 备餐台
4. Dining　　　　　4. 就餐区
5. Restroom　　　　5. 洗手间

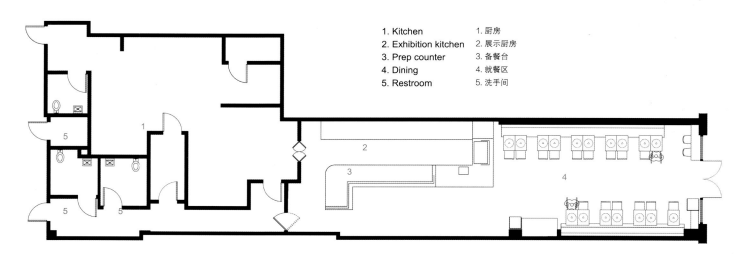

Ground Floor Plan　一层平面图

Colourful glass-clad soffit hangs above the main preparation counter　主备餐台的上方悬挂着彩色玻璃框架

Dining area incorporates seating with tall backs upholstered in a warm rust textured fabric. Above the back, on the recycled wood wall are monitors displaying the preparation methods and ingredients of the menu
就餐区的长背椅采用了柔软的衬垫和舒适的面料。椅背上方，再利用木墙上的显示屏展示着食品制作方式和新鲜的食材

Made up of two interconnecting spaces with a high ceiling, the first space is an eating area that has a parallel row of banquet seating along the walls facing an outdoor patio. The second space is the exhibition kitchen – a long cook line whose focal point is the front preparation counter displaying an array of fresh and locally-sourced natural ingredients. A backdrop colour scheme incorporating a combination of beige and taupe shades conjoins the two spatial uses.

Within the dining space, extensive use of recycled stained wood paneling has been made to create a comfortable and warm area for the dining customers. Other defining features include the use of back-painted glass and eye-catching wall graphics specially created for this venue.

餐厅由两个连锁空间组成，二者由挑高天花板连接。第一个空间是就餐区，靠墙的平行餐位正对露天平台。第二个空间是展示厨房，长条烹饪线的焦点是展示着新鲜的本地天然食材的备餐台。由米黄和灰褐两种色调交错的背景将两个空间连接起来。

在就餐区，大量回收彩色木材的运用为顾客营造了舒适温馨的空间。餐厅的其他特色体现在背涂彩色玻璃的使用和引人注目的餐厅专属墙面图案。

Detail of the main counter incorporating recycled wood components
主备餐台的细部设计，采用了回收木材

Seating banquette with monitor above　餐位及其上方的显示屏

Restaurant signage　餐厅标识

Seating wall with recycled wood paneling　再利用木板墙

Dining seating with wall graphic at rear　餐位及其后方的墙面装饰

Looking into the space through the glass door 从玻璃门看向餐厅空间

Fusao Restaurant
房雄餐厅

Completion date: 2008
Location: Lisbon, Portugal
Designer: Joao Tiago Aguiar
Photos: FG+SG Fotografia De Arquitectura
Area: 250m²

完成时间：2008年
项目地点：葡萄牙，里斯本
设计师：乔欧·迪亚戈·阿贵尔
图片版权：FG+SG建筑摄影
面积：250平方米

The programme initially given required the creation of a restaurant exclusively dedicated to Japanese cuisine inside a 5 star hotel, contemporary and innovative. The main concept of the project was to give the purity, quietness and cleanness that usually characterise this type of restaurant but at the same time with a contemporary language and using noble and luxurious materials. In that sense, the designers tried to organise the space in a clear way, and to make people feel they are in a different environment immediately after they get to the entrance.

项目要求设计师在五星级酒店中设计一家现代而创新的日式餐厅。项目的主要设计理念是在体现此类餐厅纯粹、安静、简洁的设计的同时赋予其现代设计语言和奢华的材料。设计师试图以清晰的方式组织空间，让人们一进门就有特别的感觉。

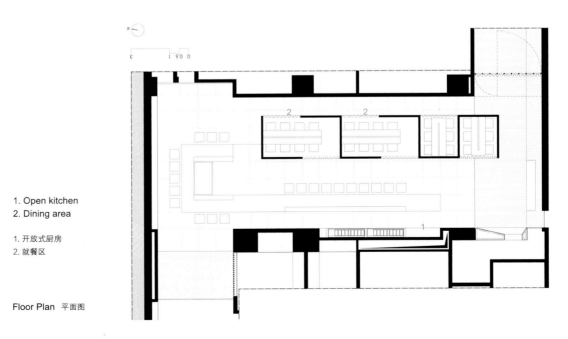

1. Open kitchen
2. Dining area

1. 开放式厨房
2. 就餐区

Floor Plan 平面图

The open-kitchen dining area becomes the focal point of the restaurant white marble counter corresponds to the concept of simpilicity and boasts elegant theme. Besides, they have also created two winter gardens at the bottom, both in small white river stone, in order to give it a more oriental atmosphere, one of them (by the entrance) being reinforced with the use of big rain forest stone panels (1.4x2.7m size) imported from the Amazonia forest in Brazil.

开放式就餐区成为了餐厅的中心——白色大理石柜台凸显简约和典雅。此外，设计师还打造了两个冬季花园，白色鹅卵石营造东方风格，而入口附近的花园采用了从巴西运来的热带雨林石板，别具特色。

The space is mainly divided by two great gestures: one is the counter made all in white marble stone, 8cm thick, that cuts the whole space and folds and turns creating different areas for diners to seat and enjoy either the Sushi and Sashimi or the Tepan-Ya-ki; the other is a great loose concrete block lifted from the ground and not touching the ceiling with a stream of light all around both below and on top that closes and opens and has all the private rooms where one can appreciate all the food in a more private, quiet and relaxed environment.

空间被划分为两大部分：一部分以8厘米厚的白色大理石台面为主，它切割了整个空间并以其蜿蜒曲折的造型划分出不同的就餐区域，供客人享用寿司、生鱼片或铁板烧；另一部分一个巨大混凝土砌块，它拔地而起，却与天花板之间留有一段距离。混凝土砌块顶部和下方都透出了光线，通过它的开合形成了若干个包房。人们可以在私密、安静、放松的包房空间中享用美食。

The ceiling was powdered by small little boxes that work as skylights, allowing the natural light to come in from various points as well.

天花板上的正方形小孔起到了天窗的作用,让自然光从各个角度照射进来。

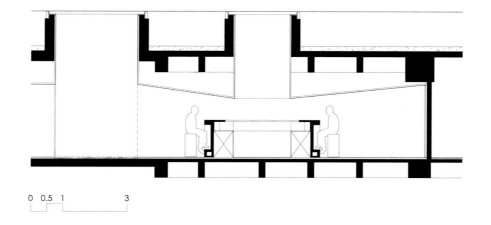

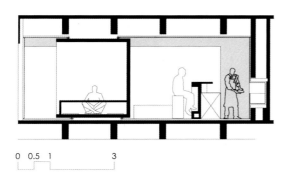

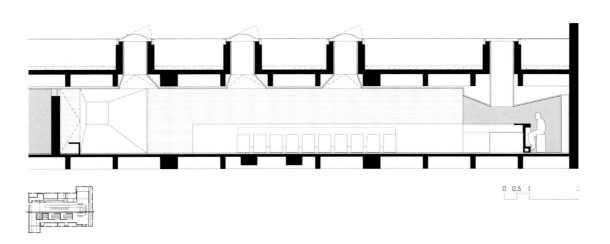
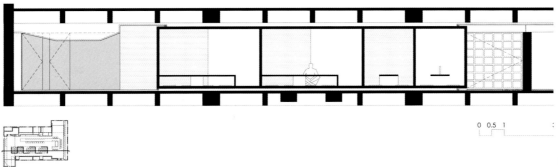

Section 剖面图

La Oliva

奥利瓦餐厅

Completion date: 2012
Location: Tokyo, Japan
Designer: Aiji Inoue, Yuki Kanai/DOYLE COLLECTION CO.,LTD.
Photos: Satoshi Umetsu/Nacasa&Partners
Area: 48m²

完成时间：2012年
项目地点：日本，东京
设计师：井上爱之、金井雄生/道尔设计公司
图片版权：梅津智/ Nacasa摄影
面积：48平方米

Before designing "La Oliva", the client and the designers visited Spain in advance. The concept for this restaurant was "to bring real Spain to Japan."

The project needed a clear concept to integrate all the below factors into one restaurant; for all of the followings are "real Spain" restaurants.

- Restaurants that emphasise bar style culture and history are most familiar throughout Spain.
- Restaurants that have modern appearance, though you can still feel the Spanish atmosphere, are relatively new type of restaurants in Spain.

Considering the construction costs, the designers wanted to make good balance of Japanese people's preference. They used contemporary methods in total and filled with combinations of beautiful town scenery and clusters of Spanish bars.

在设计奥利瓦餐厅之前，委托人和设计师先去了一趟西班牙。餐厅的设计理念是"将真正的西班牙带到日本"。

项目需要一个明确的概念将以下因素融入餐厅，以下是正宗西班牙餐厅的特征：
- 西班牙餐厅注重酒吧文化和历史
- 新式西班牙餐厅虽然外观摩登，但是仍具有浓郁的西班牙风情

考虑到建造成本，设计师希望项目能满足日本人的偏好。他们采用现代化设计方法，结合了西班牙酒吧的小城布景和美丽装饰。

Inside the restaurant, the designers placed an open kitchen so that the customers will enjoy the live atmosphere of the cooking. And the further away you sit from the open kitchen, the higher your eye level will become. In this way, no matter where you sit, everyone can easily communicate with the staff members in the kitchen.

设计师在餐厅内设计了一个开放式厨房，顾客可以享受现场厨艺表演的氛围。越远离开放式厨房的位置越高。这样一来，无论坐在那里，都能与厨房内的厨师们毫无障碍地交流。

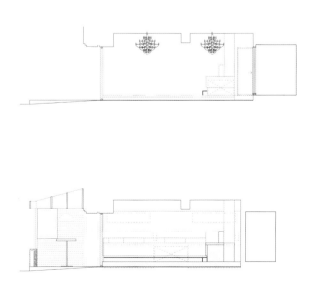

1. Entrance
2. Outdoor seating
3. Kitchen counter
4. Sofa seating
5. Floor
6. Showcase
7. Toilet
8. Kitchen

1. 入口
2. 露天餐位
3. 厨房台面
4. 沙发餐位
5. 地面
6. 展示柜
7. 洗手间
8. 厨房

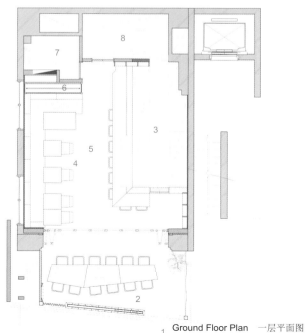

Ground Floor Plan 一层平面图

As for the space construction of the façade, the designers set terrace seats in spacious and inconspicuous boundaries with high tables to make people easier to enter the restaurant and to let the people outside know that it's always crowded.

They brought the local impression into shape; for example, bricks and walls for the blackboard, patterned imprinted tiles, and wooden counters that perfectly harmonise.

The designers displayed photographs, which they took in Spain and made it into a monochrome one, on the upper open counter as one of the designs. All the posters in this restaurant were bought from an antique shop in Spain.

对于外墙空间的设计，设计师露台餐位设在宽敞而不显眼的边界上，高大的餐桌可以方便人们进入餐厅，还可以让外面的人知道餐厅已经客座满堂。

他们将当地印象引入了设计之中；例如，以砖墙作为黑板背景，采用印花地砖和木制台面等。

设计师将一些摄于西班牙的图片单色化，然后悬挂在开放台面的上方作为装饰。餐厅内的所有海报都来自西班牙的一家古董商店。

Grand Hyatt Macau – Restaurant "MEZZA9"

澳门君悦酒店梅萨餐厅

Completion date: 2009
Location: Macau, China
Designer: Super Potato
Photos: Restaurant "MEZZA9"
Area: 1,260m²

完成时间：2009年
项目地点：中国，澳门
设计师：超级土豆工作室
图片版权：梅萨餐厅
面积：1,260平方米

Mezza9 is a restaurant on the second floor of Grand Hyatt Macau Hotel which opened in 2009 as a part of the complex facility project from a hotel and commercial facility called City of Dreams Project.

澳门君悦酒店于2009年开张，是澳门新濠天地酒店商业综合项目的一部分。梅萨餐厅就位于澳门君悦酒店的三楼。

1. Entrance
2. Private room
3. Chinese kitchen
4. Drink station
5. Terrace
6. Grill kitchen
7. Walk-in cellar
8. Dessert kitchen
9. Japanese kitchen
10. Toilet

1. 入口
2. 包房
3. 中式厨房
4. 酒水台
5. 露台
6. 烤肉厨房
7. 步入式酒窖
8. 甜点厨房
9. 日式厨房
10. 洗手间

Floor Plan 平面图

There are several show kitchens at the Asian and Western stations which allows for guests to see the cooking in front. These kitchens are deemed as the focal point of the restaurant, which not only present the mysterious cooking process to diners, but also add vividness to the entire restaurant. Kitchens' design varies in form and function according to the food provided.

餐厅内设有多个亚洲和西方开放式厨房，供宾客观看厨艺表演。这些厨房被视为餐厅的焦点，不仅向食客们展示了神秘的烹饪过程，还为整个餐厅增加了活力。根据其所提供的菜品，各个厨房的设计在形式和功能上各有不同。

The interior is expressed in a modern manner through materials which elaborates naturalism in a strong way such as stones which were cut out from the mountains, timber with natural faces left, and iron with rust. By segregating these materials with usage of floor, wall and counter, the materials do not only become a material for the interior, yet, also elements to construct an art installation space.

Created by cutting-edge Japanese interior design firm Super Potato, this eclectic, highly textured 292-seat space is defined by giant, roughly hewn stone blocks forming counters and lining walls, as well as an eye-catching array of lattice-patterned metal screens. Besides the main dining room, guests can choose the outdoor terrace, one of four semi-private, Japanese wood-lined booths, or one of two 10- to 12-seat private rooms.

室内设计通过自然材料（如来自高山的石材、裸露自然表面的木材和生锈的钢铁）的运用呈现出现代感。这些材料在地面、墙壁和台面上的运用使其不仅成为了室内装饰材料的一部分，还成为了构建艺术空间的重要元素。

餐厅由日本先锋设计公司超级土豆进行设计，可容纳292人同时就餐。餐厅以巨大而粗犷的石块形成了台面和墙围，网格图案的金属屏风同样引人注目。除了主餐室之外，宾客还可以选择露天平台、半私密包房、日式木包间、私人包房等处进行就餐。

barQue

巴Q餐厅

Completion date: 2011
Location: Athens, Greece
Designer: k-studio
Branding: DGGD, k-studio
Lighting: Halo
Photos: Vangelis Paterakis

完成时间：2011年
地点：希腊，雅典
设计：k工作室
品牌设计：DGGD公司、k工作室
灯光设计：Halo照明设计公司
图片版权：范吉利斯·帕特拉基斯

The cuisine of barQue revolves around barbecued meats. This was the starting point for the design and the branding concept. In collaboration with DGGD the branding completes the atmosphere by borrowing the font from the tool used to brand meat before hanging, to design the restaurant logo. The "Taste the Fun" neon motto reminds everyone that barbequing is a sociable activity and that the preparation, grilling and eating of good quality meat should be celebrated.

巴Q餐厅的美食围绕着各式烤肉展开，这也是设计和品牌理念的出发点。k工作室和DGGD公司在品牌设计中引用了鲜肉贩售标牌来作为餐厅的标识。"品尝趣味"字样的霓虹灯提醒人们烤肉是一项社交活动，准备、烧烤和食用高品质肉类是一件美好的事情。

Barbequing is a social activity, so the design opens up the kitchen and allows the choreography of the chefs to become part of the dining experience, with some diners even sitting along the kitchen worktop to eat. A steel grille, reminiscent of a barbeque grill, acts as a false ceiling that accentuates the height and gives the space a dark, masculine weight.

A deep frieze of cut pine blocks, arranged as an abstraction of the traditional butchers chopping table, is suspended above the bar allowing continuity between the interior and exterior dining areas.

The glow of the wood-block frieze combines with varied pieces of wooden furniture and a large, glass-jar chandelier to warm the space, balancing soft, crafted and delicate textures with the rough black steel ceiling and the blaze of the grill from the open kitchen.

烤肉是一项社交活动，因此设计师设计了一个开放式厨房，让厨师的技艺成为就餐体验的一部分。一些顾客甚至坐在厨房台面上就餐。设计师以象征着烤肉架的钢制网作为吊顶，突出了空间的高度，并增加了深邃的阳刚之感。

松木砖块被排列成传统切肉板的造型，悬挂在吧台上方，将室内外就餐区连接了起来。

木砖块与各式木家具相结合，在大型玻璃瓶吊灯的照映下传递出温暖的光线，平衡了柔和、精巧的材质与粗犷的黑色天花板和开放式厨房的烤肉火光。

1. Open kitchen
2. Dining area

1. 开放式厨房
2. 就餐区

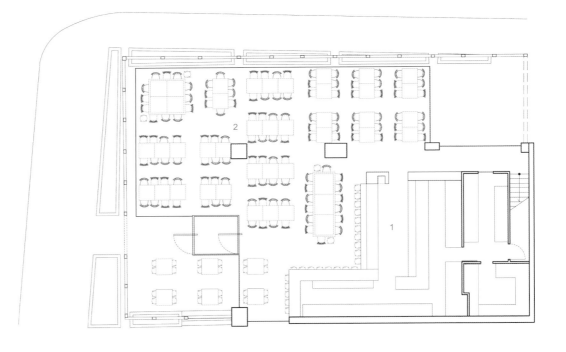

Ground Floor Plan 一层平面图

La Nonna Restaurant

拉诺那餐厅

Completion date: 2010
Location: Mexico City, Mexico
Designer: Cherem Serrano
Photos: Jaime Navarro
Area: 275m²

完成时间：2010年
项目地点：墨西哥，墨西哥城
设计：切利姆·塞拉诺
摄影：杰米·纳瓦罗
面积：275平方米

La Nonna is an Italian restaurant located in La Condesa. This project was made in collaboration with DMG architects. The design premise was to liberate the plan so the restaurant could take the most advantage of the 200 square metres local. The restaurant fuses its environment with simplicity.

拉诺那是一家位于拉肯德撒的意大利餐厅，由 DMG 建筑事务所联合设计。设计要求解放规划，让餐厅充分利用 200 平方米的空间。餐厅以简洁朴素的设计与周边环境结合在一起。

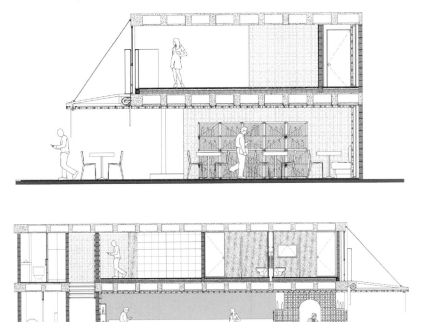

Sections 剖面图

The mirror was placed in order to enlarge the space and create a game of light and shadows. The furniture is elegant and simple. At the centre is placed a bar with a pizza stove and a wood counter to sit and enjoy a nice wine.

Incorporating the use of local materials, the floor is made of dark stone and is built unto the sidewalk, while the walls and ceiling are surrounded by red brick with special cutting on top of mirrors.

The services area in a second level connected with the ground floor by an elegant illuminated staircase. While going down to the main level, the architects created a playful mezzanine that catches the eye of the visitor by letting light in and setting a swinging chair, some books and a floor lamp. The ceiling in the stair and main circulations is painted in black in order to emphasise the design of the red bricks placed in the restaurant area.

镜子的设置扩大了空间，形成了光影的交错。家具优雅而简约。餐厅中央的吧台内设有比萨炉，人们可以在木台上享用美酒。

地板设计融合了当地材料，由黑石铺成，一直延伸到人行道；而墙壁和天花板则由红砖包围，在镜子上方进行了特殊的切割。

二楼服务区通过优雅的发光楼梯与一楼相连。建筑师打造了一个趣味化的夹层楼，以渗透的光线、摇椅、书籍和落地灯吸引眼球。楼梯和主要通道的天花板被漆成了黑色，突出了餐厅区域的红色砖石设计。

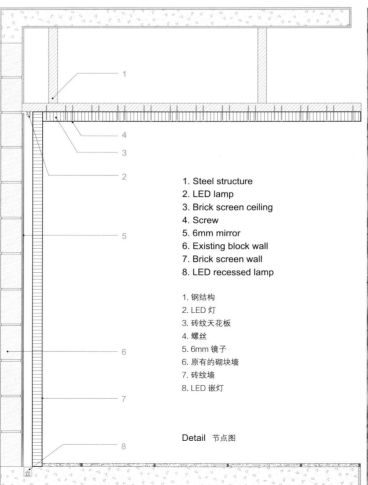

1. Steel structure
2. LED lamp
3. Brick screen ceiling
4. Screw
5. 6mm mirror
6. Existing block wall
7. Brick screen wall
8. LED recessed lamp

1. 钢结构
2. LED 灯
3. 砖纹天花板
4. 螺丝
5. 6mm 镜子
6. 原有的砌块墙
7. 砖纹墙
8. LED 嵌灯

Detail 节点图

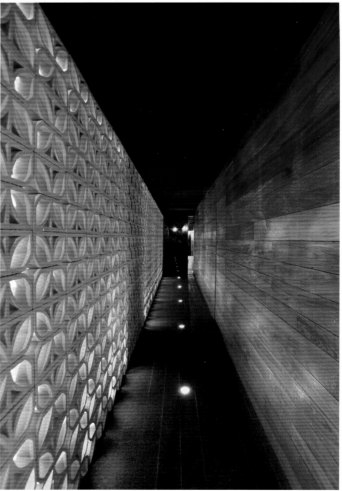

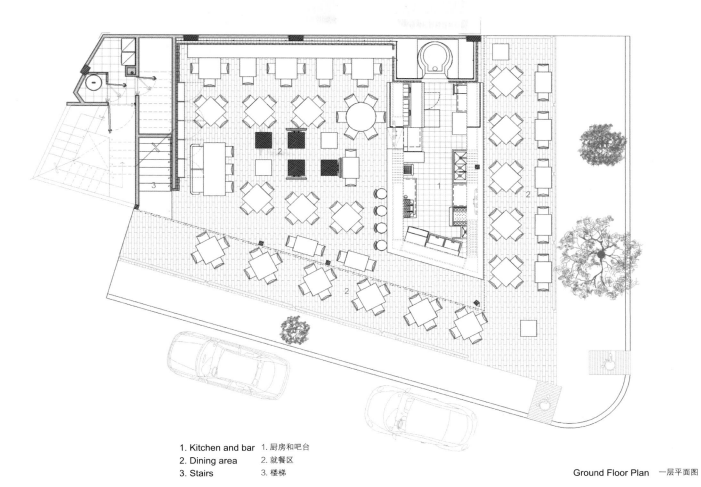

1. Kitchen and bar　1. 厨房和吧台
2. Dining area　2. 就餐区
3. Stairs　3. 楼梯

Ground Floor Plan　一层平面图

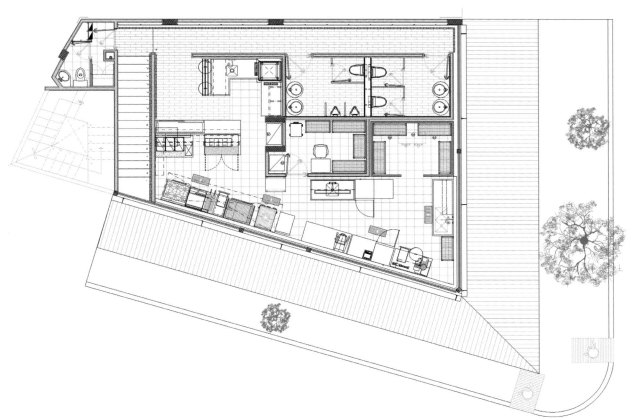

First Floor Plan　二层平面图

The illumination is hidden within the bricks and creates a nice rhythm of light that accentuates the architects concern to integrate the architecture with the functional requirements.

照明隐藏在红砖后面，营造出良好的韵律感，体现了建筑师将建筑与功能相结合的设计理念。

Jaffa – Tel Aviv Restaurant

特拉维夫雅法餐厅

Completion date: 2011
Location: Tel Aviv, Israel
Designer: Alon Baranowitz and Irene Kronenberg
Photos: Amit Geron
Area: 350m²
Award: 3rd Prize at the AIT Award 2012/ Category: Bar and Restaurants

完成时间：2011年
项目地点：以色列，特拉维夫
设计：阿龙·巴拉诺维奇与艾琳·克隆恩伯格
图片版权：阿密特·格龙
面积：350平方米
获奖情况：2012年AIT大奖酒吧餐厅类三等奖

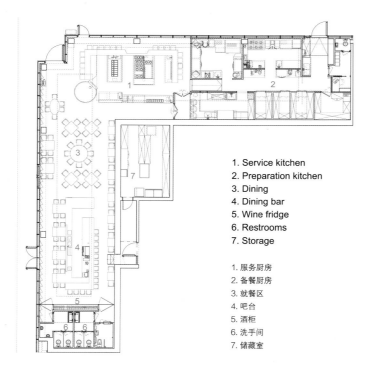

1. Service kitchen
2. Preparation kitchen
3. Dining
4. Dining bar
5. Wine fridge
6. Restrooms
7. Storage

1. 服务厨房
2. 备餐厨房
3. 就餐区
4. 吧台
5. 酒柜
6. 洗手间
7. 储藏室

Floor Plan 平面图

Haim Cohen, widely known as Israel's first celebrity chef, appointed BK Architects for the design of his new restaurant in Tel Aviv: "Jaffa – Tel Aviv".

In Israel many culinary traditions live together. Inspired by their richness, Cohen considers them only as a starting point. With a touch of ingenuity, bearing honesty and subtle simplicity, he brings them to an utterly fresh and inventive level of cuisine. Cohen reinvents the Israeli cuisine, he challenges its tradition on a daily basis, yet he always pays homage to its roots.

BK Architects started off at the same point as Haim with an aim to represent the spirit of his kitchen. Using simplicity and honesty as primer building materials, they make use of raw and ordinary finishes. However, they are only their starting point.

以色列的第一位名厨海姆·科恩委托BK建筑事务所对他的新餐厅"特拉维夫雅法餐厅"进行设计。

以色列拥有众多烹饪传统。科恩受到它们的启发，开始了自己的烹饪追求。他以独创、真诚和简约的手法将传统烹饪带入了一个全新的时代。科恩重新创造了以色列美食，他挑战传统，同时也对其充满敬意。

BK建筑事务所的设计与海姆·科恩的出发点相同，旨在体现他的厨房精神。他们运用简约实在的建筑材料，进行了简单的装饰。但是，这仅仅是一个起点。

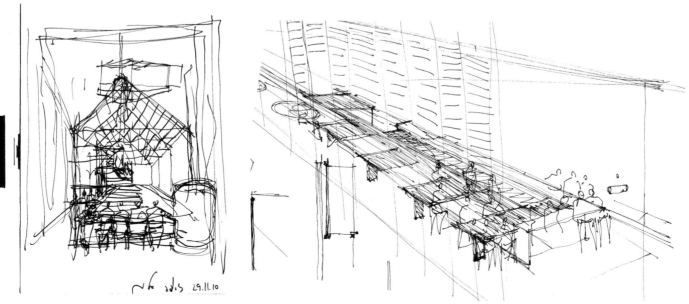

The open kitchen in Jaffa – Tel Aviv is the embodiment of "hospitality". Guests are invited to seat around the burning fire and watch the "show". Everything is within reach, while nothing is left behind the stage. The kitchen's stainless steel tops become a huge dining bar.

特拉维夫雅法餐厅的开放式厨房是热情好客的化身。客人可以围坐在炉火周围观赏大厨的表演。一切都触手可及，没有什么是隐藏起来的。厨房的不锈钢台面变成了巨大的吧台。

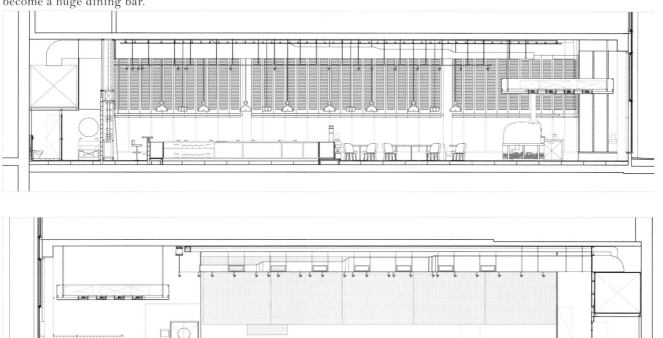

Sections 剖面图

The restaurant envelope is done in the very basic materials: water, cement, aggregates and steel. Exactly as Haim's cuisine: water, flour and olive oil. The exposed concrete walls of the envelop were polished to reveal the true nature of the stone aggregates, while the concrete floor and ceiling were left untouched.

Like the culinary ingredients which come from different lands, many design elements are mixed together.

The patch work of Turkish carpets hung along the restaurant walls represents a cultural kaleidoscope. Each carpet piece has a story, an identity and a home land. They all evoke the flair of old Jaffa, a city where multi-national traditions live side by side.

The tall poplar wood shutters facing the west façade allow the sun to play a magical symphony of light and shadow upon the interior walls, a play one can always find walking the alleys of Old Jaffa. The Piet Hein Eek scrap-wood chairs and tables together with second-hand furniture pieces and outdoor light "bells" from the Czech Republic, bestow sophisticated representation of a cultural melting pot. The poured terrazzo of the alcohol bar, and the raw metal winery make a destination zone away from the lively kitchen.

餐厅的外观设计采用了最基本的材料：水、水泥、聚合物和钢。这与海姆·科恩的烹饪食材有异曲同工之妙：水、面粉和橄榄油。露石混凝土外墙经过抛光，展现了石头颗粒的自然特征，而水泥地面和天花板则未经处理。

正如烹饪食材来自世界各地，许多设计元素也被混合在一起。

餐厅墙面上的土耳其挂毯体现了文化的多样化。每块挂毯都有一个故事、一种身份和一个故乡。它们唤起了人们对雅法古城——一座融合了多民族传统的古城的记忆。

西面的杨木百叶窗让阳光在内墙上留下了奇妙的光影效果，正如人们在雅法古城的小巷中所看到的一样。由著名设计师海恩·伊克所设计的桌椅与二手家具、来自捷克的露天铃铛一起，创造了文化熔炉的效果。水磨石酒吧吧台和粗金属酒架是厨房外的另一处风景。

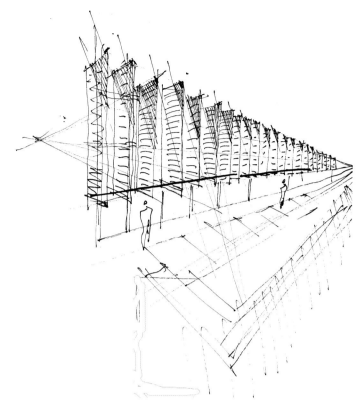

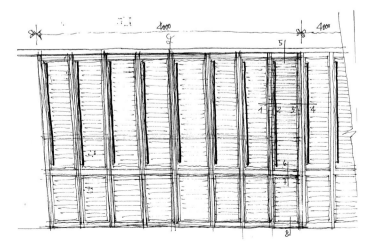

Aoyama Hyotei

青山冰帝餐厅

Completion date: 2009
Location: Tokyo, Japan
Designer: PODA, asterisk studio
Photos: Takumi Ota
Area: 216m²

建成时间：2009年
地点：日本，东京
设计师：PODA星号工作室
图片版权：太田拓实
面积：216平方米

Japanese restaurant "Aoyama Hyotei" is located on the first basement floor of a multi-tenant building along Aoyama Dori, the main street running across one of the most sophisticated shopping and business hubs in Tokyo.

青山大道是横跨东京最繁华的商业中心的主要街道，青山冰帝日本餐厅就位于青山大道旁一座多户住宅楼的地下室里。

1. Entrance
2. Entrance hall
3. Cloak
4. Bar
5. Open kitchen
6. Counter seating
7. Bench
8. Gingko objet
9. Private room
10. Washing room
11. Lavatory

1. 入口
2. 门厅
3. 衣帽间
4. 吧台
5. 开放式厨房
6. 吧台座位
7. 长椅
8. 银杏面板
9. 包房
10. 洗碗间
11. 洗手间

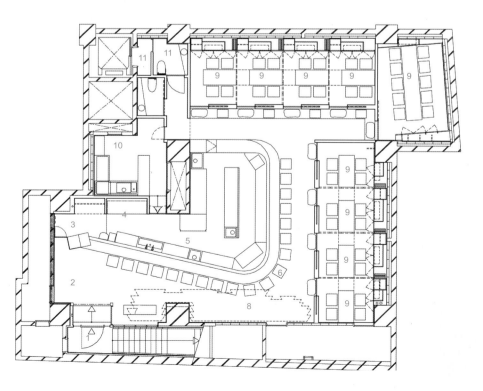

Floor Plan 平面图

The restaurant features a long white counter winding around a stainless steel kitchen, and two rows of private rooms behind the walls finished with metallic brown tiles. The former countertop made of white synthetic stone shows the elegant colour of the traditional earthenware to advantage. Enjoying the dynamic cooking performed over a charcoal fire and steaming pots, the diners will be intimate spectators to share a long white dish.

Private rooms are mainly prepared for the stylish neighbours and celebrities who host a business reception or a private banquet. The ceiling of a private room painted in gold is formed in two house forms crossing. A row of four private rooms can also be used as a continuous party room, while the partitions between the rooms are stored inside the closets.

A passage between a counter and private rooms, whose ceiling is the highest, represents an open air garden alley to invite guests into tea houses. An illuminated object made of reinforced paper is raised besides the entrance hall, which is resulted from the abstraction of the stunning rows of ginkgo trees nearby. The illumination is programmed to change its colour automatically, and delight the eyes of the guests. Ginkgo leaves turning gold yellow blown from the outside, shining proudly on the sprinkled garden alley, are represented with scattered sheets of gold foils on the dark mortar floor, sealed with seamless shiny resin. The abstraction of each seasonal element is valuable not only to make dishes lively, but to take their mind off of the fact that they are dining in the basement.

餐厅的开放式不锈钢厨房外围环绕着一圈白色台面，两排包房则隐藏在金属棕色瓷砖装饰的墙后。开放式厨房的吧台由白色人造石制成，凸显了传统陶器的优雅色彩。炭火和蒸锅上动感十足的厨艺表演拉近了就餐者与餐厅之间的距离。

包房主要供时尚的友邻和名流举办商务接待或私人宴会。包房的天花板被涂成了金色，呈现为两座屋顶交叉的形态。可以将列成一排的四间包房的隔断移开，储藏到壁柜中，从而形成一个连续的派对空间。

吧台和包房之间走道的天花板最高，走道象征着邀请顾客进入茶室的露天花园小巷。门厅上方悬挂着由增强纸制成的照明面板，呼应着附近令人惊艳的银杏树。面板会自动变换色彩，令顾客眼前一亮。外面的银杏叶变成金黄，随风飘落在花园小巷上；餐厅内则以粘贴在黑色灰泥地板上金箔遥相呼应。抽象的季节元素不仅让菜肴变得鲜活，还让客人忘记了自己置身于地下室就餐的事实。

Esquire
君子餐厅

Completion date: 2011
Location: Brisbane, Australia
Designer: HASSELL
Photos: Roger D'Souza
Area: 425m²

完成时间：2011年
项目地点：澳大利亚，布里斯班
设计师：HASSELL设计公司
图片版权：罗杰·德索萨
面积：425平方米

This project was conceived and delivered through the close collaboration of architecture and interior design disciplines. An art curator was also engaged in the design process and remains responsible for the seasonal rotation of artworks. HASSELL worked very closely with the restaurant owners throughout the project to achieve a very high standard of materiality and workmanship for cost that was within their budget. The design also pursues a sustainability agenda of building well and building once through robust design and natural, durable materials that will not need to be replaced due to functional or visual decay over the life of the restaurant's operation, which is at least 10 years.

Esquire's interior design of a series of spatially diverse and connected rooms offers an experience for everyone to enjoy.

Transformed from a large commercial space into a series of intimate dining areas, guests can stop in for lunch, take pleasure in the restaurant's fine dining menu or enjoy a drink at the bar.

项目的设计和建造严格遵循了建筑与室内设计法则。餐厅还专门请来了一位专业展览设计人士对艺术品的摆放进行了指导。HASSELL设计公司在项目过程中与餐厅业主紧密合作，在有限的预算中实现了高品质材料和手工的运用。项目所采用的稳健设计和天然、耐久材料实现了建筑的可持续化，近10年内，餐厅无需为材料的功能和视觉效果衰退而进行翻新。

君子餐厅的室内设计由一系列不同的空间和相互连接的房间组成，为人提供了愉悦的体验。

餐厅将大型商业空间改造成多个私密的就餐区，顾客可以在酒店享用午餐美食，或在吧台边品味一杯美酒。

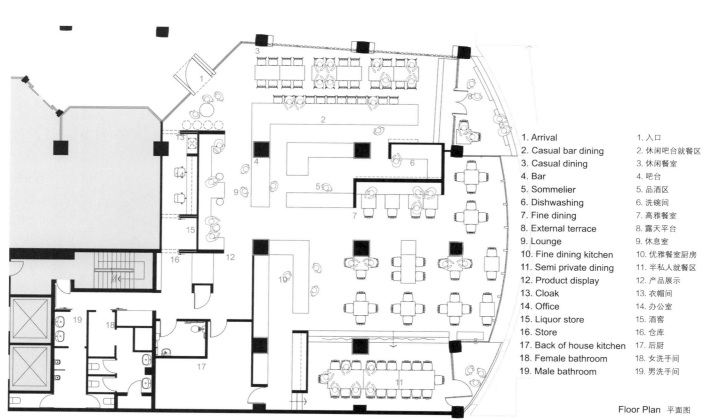

1. Arrival	1. 入口
2. Casual bar dining	2. 休闲吧台就餐区
3. Casual dining	3. 休闲餐室
4. Bar	4. 吧台
5. Sommelier	5. 品酒区
6. Dishwashing	6. 洗碗间
7. Fine dining	7. 高雅餐室
8. External terrace	8. 露天平台
9. Lounge	9. 休息室
10. Fine dining kitchen	10. 优雅餐室厨房
11. Semi private dining	11. 半私人就餐区
12. Product display	12. 产品展示
13. Cloak	13. 衣帽间
14. Office	14. 办公室
15. Liquor store	15. 酒窖
16. Store	16. 仓库
17. Back of house kitchen	17. 后厨
18. Female bathroom	18. 女洗手间
19. Male bathroom	19. 男洗手间

Floor Plan 平面图

Striking a careful balance between loft and luxury, ceiling-high windows elegantly frame Brisbane's Story Bridge, while the open plan kitchens put diners in direct contact with the chefs as they prepare and plate each dish.

Through the design, the HASSELL team set out to complement the artistic presentation of the food by award-winning chef Ryan Squires and his business partner, Cameron Murchison. Traditional approaches have been reinterpreted into a design that emphasises natural materials and concise detailing.

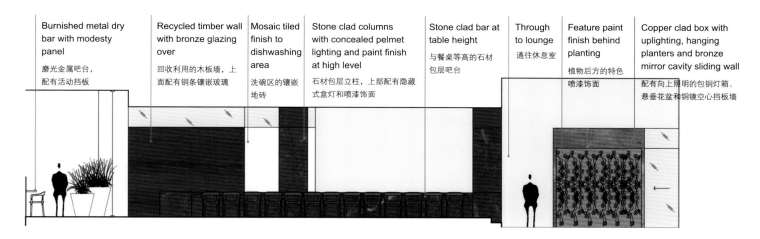

View across casual dining towards bar　从就餐区看向吧台

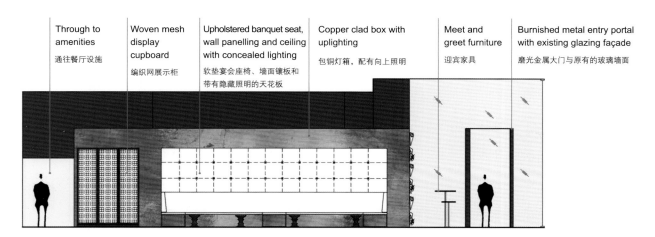

View across entry and lounge towards amenities corridor　从入口和休息室看向餐厅走廊

餐厅的落地窗既优雅又奢华，从窗口望去是布里斯班故事桥的美丽景色。开放式厨房则让顾客可以与工作中的厨师直接交流。

在设计中，HASSELL 团队力求展现著名厨师赖安·斯奎尔斯和他的搭档卡梅伦·默奇森的精湛技艺。设计重新诠释了传统的天然材料和简洁的细部设计。

The elevated private dining room features a long wall layered with heritage timber mouldings, which are washed with light and cast into deep, textural relief.

Blown glass ball lights are suspended from Esquire's ceiling, which rises and falls in accordance with each dining space to produce a subtle but tangible variation in size and ambience.

抬高的私人包房的墙壁上层叠着复古木墙板，墙面在灯光的照射下显示出深邃的纹理。

餐厅的天花板上悬挂着吹制玻璃制成的球形灯。天花板随着就餐空间的变化而起伏，在规模和氛围上形成了一种微妙的变化。

Chapter 5: Fire Prevention in Open Kitchens
第五章：开放式厨房的防火

5.1 Fireproof Units and Partitions

5.1 防火单元及分隔

5.2 Safe Evacuation

5.2 安全疏散

5.3 Liquefied Petroleum Gas Pot Storage

5.3 液化石油气瓶库

5.4 Firefighting Equipment

5.4 消防设施

5.4.1 Automatic Fire Alarm System

5.4.1 火灾自动报警系统

5.4.2 Automatic Fire Extinguishing System

5.4.2 自动灭火系统

5.4.3 Smoke Exhaust System

5.4.3 排烟系统

5.4.4 Kitchen Firefighting Equipment

5.4.4 厨房灭火设备及器材

5.4.5 Mechanical Air Supply and Exhaust System

5.4.5 机械送排风系统

5.4.6 Evacuation Signs and Emergency Lighting

5.4.6 疏散指示标志和应急照明

Chapter 5: Fire Prevention in Open Kitchens

According to relative statistic data, 80% of the fire in dining places originates from kitchens. As a main location for fire, the fire prevention of the kitchen is especially important.

Fire Prevention Solutions of Open Kitchens
Due to the high potential of fire in the dining place with an open kitchen, it is inappropriate to locate it in a building with fire resistance rating lower than three. It should be located in buildings with first or second rate fire resistance rating.

5.1 Fireproof Units and Partitions

No matter the dining place with an open kitchen is located in a high-rise, multi-storey or single-storey building, it should be considered as an individual fireproof unit. First of all, the relevant components' of fire resistance should meet the design requirements. Secondly, the partition design of these places can refer to the requirements of entertainment places: to use partitions with fire resistance rate higher than 2.0h to separate it from other parts and to install first or second rate fireproof gate with thermal insulation. The area of each unit varies according to different building, storey and location and should conform to relevant rules and requirements. If the kitchen and dining area is separated by non-loading glass partition, the glass partition's fire resistance rate should achieve first rate.

5.2 Safe Evacuation

The place where open fire is used should locate near the exterior wall and try to keep away from evacuation route and exit. The kitchen and dining area should possess respective evacuation exit. At least two exits should be set near the open kitchen or the kitchen is connected to dining area with evacuation route. On the premise of at least two exits in dining area, the exits should be located far from open fire and in different directions. The technical specifications of evacuation distance can refer to those of other public spaces and be improveed according to needs.

相关统计数据表明，在餐饮场所发生火灾中，大约80%起源于厨房。厨房作为公共场所火灾发生的主要部位和多发部位，它的防火安全显得尤为重要。

开放式厨房的防火设计解决措施
由于开放式厨房的餐饮场所火灾危险性大，故不宜设置在耐火等级三级以下的建筑物内，宜设置在耐火等级一、二级的建筑物内。

5.1 防火单元及分隔

以开放式厨房为经营模式的餐饮场所不论设置在高层建筑还是多层建筑或者单层建筑内，均应视为一个独立的防火单元来进行考虑。首先，建筑的耐火等级相应的构件应满足设计要求；其次，此类场所的分隔可参考歌舞娱乐放映游艺场所的要求，采用耐火极限不低于2.0h的隔墙与其他部位隔开，当墙上开门时应设置具有隔热功能的甲级或乙级防火门；每个单元的面积根据设置的建筑、楼层、部位的不同，防火分区的面积应符合相应规范要求。若厨房与餐厅之间采用非承重玻璃隔墙进行分隔，宜采用耐火性能等级为I级的防火玻璃隔墙。

5.2 安全疏散

使用明火的部位，应尽量靠外墙设置，尽量避开安全疏散走道和出口部位。厨房与就餐区的疏散出口应分别设置，在开放式厨房部位应设置不少于2个疏散出口，或是在厨房内应有环通的疏散走道与餐厅区连通；就餐区的疏散出口在不少于2个的前提下，疏散出口的设置应避免离明火区域太近，且应设置在不同方向；对于疏散距离的技术指标可对应其他公共场所的疏散指标予以提高。

5.3 Liquefied Petroleum Gas Pot Storage

The places where liquefied petroleum gas pots are used should set a separate pot storage, which should not be located in the middle of a storey or middle storey. The doors and windows of the storage should open outwards, never to the kitchen or inwards. The storage should have necessary explosion-proof pressure relief surface, which is kept away from public gathering place and main circulation. The storage should include an inflammable gas detector which connects to a 24h duty room. All the electrical equipment should be explosion-proof. It is prohibited to use liquefied petroleum gas pots in high-rise civil buildings, public entertainment places and underground spaces.

5.4 Firefighting Equipment

Because the decoration of dining place typically builds on existing building and the fireproof partitions exceed existing design requirements, it is necessary to enhance safety through the installation of firefighting equipment in interior design. The configuration of firefighting equipment can refer to those required in entertainment places, including automatic fire alarm system, sprinkler system, smoke control and exhaust system, kitchen fire extinguishing equipment, evacuation signs and emergency lighting.

5.4.1 Automatic Fire Alarm System
·The selection of detector. The places where pipelined gas, natural gas, gas meter and liquefied petroleum gas are used should select inflammable gas detector; the kitchen should select thermal detector; the dining area should select smoke detector. The places equipped with interactive devices and automatic fire extinguishing system should use the combination mode of smoke detector, thermal detector and inflammable gas detector.
·System interaction function. The alarm signal of detector is feedback to the control centre, and leads to shut off the emergency automatic stop valve of gas pipeline, to activate accident exhaust system, mechanical

5.3 液化石油气瓶库

使用瓶装液化石油气的场所应设置独立的瓶库，且瓶库不得设置在楼层中间的位置和中间楼层。瓶库的门、窗应直接向外开启，不得向厨房或建筑内开启。瓶库应设置必要的防爆泄压面，泄压面应避开公众聚集场所和主要交通道路。设置可燃性气体探测装置，报警信号须接至昼夜有人值班场所。瓶库内电气设备应为防爆型。严禁在高层民用建筑、公共娱乐场所、地下空间内使用瓶装液化石油气。

5.4 消防设施

由于餐饮场所的装修一般在建筑原有的基础上进行，且防火分隔上大大突破现有设计规范的要求，因此通过后期的装修设计中加强消防设施的设置来提高场所的安全性必不可少。消防设施的设置，可参照设计规范中对歌舞娱乐放映游艺场所的消防设施设置要求进行设计，包括火灾自动报警系统、自动喷水灭火系统、防排烟系统、厨房灭火设备及器材、疏散指示标志及照明灯。

5.4.1 火灾自动报警系统
a. 探测器的选择。使用管道煤气、天然气、燃气表房及存储液化石油气的场所宜选择可燃气体探测器，厨房部位宜选择感温探测器，而在餐厅区宜选择感烟探测器。对于装有联动装置、自动灭火系统的场所，宜采用感烟探测器、感温探测器、可燃气体探测器的组合形式。

Chapter 5: Fire Prevention in Open Kitchens

smoke control and exhaust system, kitchen self-extinguishing devices, sprinkler system and emergency broadcasting system and to shut down non-fire power, access control system and normally open gate.

·Simple alarm system. For the smaller place inside a building without an alarm system, single point automatic fire alarm is an option. For example, a free standing inflammable detector can be activated to shut off gas pipeline valve and to open kitchen firefighting equipment or accident exhaust.

5.4.2 Automatic Fire Extinguishing System

For the building with a sprinkler system, the interior should be protected. The building without a sprinkler system should install simple fire-sprinkling system with quick response sprinkler heads. Automatic fire extinguishing devices should be installed in exhaust hood and cooking parts.

5.4.3 Smoke Exhaust System

The smoke exhaust of open kitchen should be separated from others. When natural smoke extraction is used, the window area should be maximised. In other words, the window area should not be less than 5% of the unit's area. When natural smoke extraction is not possible, a mechanic exhaust system should be installed. When in charge of one unit, the smoke exhaust rate should achieve 60m^3/h; when in charge of two or more units, the rate should not be less than 120m^3/h. No matter how large the open kitchen is, the connections with other parts should install smoke curtain.

5.4.4 Kitchen Firefighting Equipment

According to relevant provisions, it is recommended to install self-extinguishing equipment in fume exhaust cover and cooking area. The kitchen could also be equipped with portable kitchen extinguishers and fire blankets.

5.4.5 Mechanical Air Supply and Exhaust System

A free standing mechanical air supply and exhaust system should be installed in the kitchen. In normal working condition, the air change rate

b. 系统联动功能。探测器在发出报警信号后应将信号反馈至控制中心，并联动关闭燃气管道上的紧急自动切断阀，启动事故排风系统、机械防排烟系统、厨房自动灭火装置、自动喷水灭火系统、应急广播系统、关闭非消防电源、门禁系统断电释放、常开防火门相应关闭等。

c. 简易报警系统。对于建筑内无报警系统且面积不大的场所，可采用单点式火灾自动报警装置。如采用独立式可燃气体探测器，触发时联动关闭燃气管道阀门、启动厨房灭火设备或开启事故排风等。

5.4.2 自动灭火系统

建筑内有自动喷水灭火系统的，应对场所内全保护；若建筑物内无自动喷水系统的，应设置简易喷水灭火系统；洒水喷头宜选择快速响应喷头。烹饪操作间的排烟罩及烹饪部位应设置厨房自动灭火装置。

5.4.3 排烟系统

开放式厨房与其他部位的排烟分开设置。采用自然排烟方式时，场所内的开窗面积宜取设计规范中的最大值，也就是可开启的外窗面积不应小于该独立的单元面积5%；无自然排烟条件的，应设置机械排烟系统。当排烟机担任一个防烟分区排烟时，每平方米的排烟量按60m^3/h计算，担负两个或两个以上防烟分区排烟时，每平方米按不小于120m^3/h计算。同时无论开放式厨房面积多大，与其他部位连接处宜设置排烟垂壁。

5.4.4 厨房灭火设备及器材

根据相关规定，建议开放式厨房无论面积大小，在排油烟罩及烹饪部位均设置自动灭火装置；厨房内可适量配置手提式厨房专用灭火器、灭火毯等灭火器材。

should not be less than 6 per hour. In accident condition, the air change rate should not be less than 12 per hour.

5.4.6 Evacuation Signs and Emergency Lighting
Smaller places could conform to existing design standards. However, large places such as large food court with complicated interior layout should add continuous lighting evacuation signs on evacuation walk and main evacuation route besides business areas, evacuation ways and exits.

The emergency lighting in business area, evacuation walk and stairs could apply unified illuminance. For instance, the minimum ground illuminance of the lights in the staircase should achieve 5.0 lx with appropriate spacing.

5.4.5 机械送排风系统
应设置独立的机械送排风系统，在正常工作状态，系统的通风换气次数不应小于 6 次 /h，当转换为事故通风状态时，换气次数不应小于 12 次 /h。

5.4.6 疏散指示标志和应急照明
对于面积相对较小的场所，可按照现行设计规范进行。但对于面积大的场所，如大型美食广场等，考虑到内部格局的复杂，除在营业区、疏散通道、安全出口处设置疏散指示标志和完全出口标志外，在疏散走道和主要疏散路线的地面上或靠近地面的墙上宜附加设置连续的灯管疏散指示标志。

在营业厅、疏散走道、疏散楼梯等部位设置的应急照明设施，其照度可以进行统一，如均采用楼梯间内的地面最低照度不低于 5.0 lx 的指标，且设置的间距不宜太大。

Index 索引

AkarStudios
Add: Santa Monica, USA
Web: www.akarstudios.com

Alon Baranowitz and Irene Kronenberg
Add: Tel Aviv, Israel
Web: www.bkarc.com

Arch. Camilla Lapucci & Lapo Bianchi Luci The Studio CL
Add: Italy

asterisk studio
Add: San Francisco, USA
Web: www.studioasterisk.com

Cetra Ruddy
Add: New York, USA
Web: www.cetraruddy.com

Cherem Serrano
Add: Distrito Federal, México
Web: www.cheremarquitectos.com

Chiho&Partners
Add: Seoul, Korea
Web: www.chihop.com

Concrete Architectural Associates
Add: Amsterdam, The Netherlands
Web: www.concreteamsterdam.nl

D/Dock
Add: Amsterdam, The Netherlands
Web: www.ddock.nl

DOYLE COLLECTION CO.,LTD.
Add: Tokyo, Japan
Web: www.doylecollection.jp

Gilles & Boissier
Add: Paris, France
Web: www.gillesetboissier.com

HASSELL
Add: Austrialia
Web: www.hassellstudio.com

João Tiago Aguiar
Add: Lisbon, Spain
Web: www.joaotiagoaguiar.com

Josep Ferrando
Add: Barcelona, Spain
Web: www.josepferrando.com

k-studio
Add: Athens, Greece
Web: www.k-studio.gr

Landini Associates Pty Ltd
Add: Sydney, Australia
Web: www.landiniassociates.com

Maura
Web: www.mauradesign.com

Minas Kosmidis
Web: www.minaskosmidis.com

Stephen Williams Associates
Add: Hamburg, Germany
Web: www.stephenwilliams.com

Super Potato
Add: Tokyo, Japan
Web: www.superpotato.jp

Waraphan Watanakaroon
Add: Bangkok, Thailand
Web: www.slideshare.net

图书在版编目（CIP）数据

开放式餐厅 / （荷）沃尔托编；常文心译. -- 沈阳：辽宁科学技术出版社，2014.3
　ISBN 978-7-5381-8442-6

Ⅰ. ①开… Ⅱ. ①沃… ②常… Ⅲ. ①餐厅－室内装饰设计－世界－图集 Ⅳ. ①TU238-64

中国版本图书馆CIP数据核字(2013)第318509号

出版发行：辽宁科学技术出版社
　　　　　（地址：沈阳市和平区十一纬路29号 邮编：110003）
印　刷　者：利丰雅高印刷（深圳）有限公司
经　销　者：各地新华书店
幅面尺寸：215mm×285mm
印　　张：15
插　　页：4
字　　数：50千字
印　　数：1～1200
出版时间：2014年 3 月第 1 版
印刷时间：2014年 3 月第 1 次印刷
责任编辑：陈慈良　鄢　格
封面设计：何　萍
版式设计：何　萍
责任校对：周　文
书　　号：ISBN 978-7-5381-8442-6
定　　价：228.00元

联系电话：024-23284360
邮购热线：024-23284502
E-mail：lnkjc@126.com
http://www.lnkj.com.cn
本书网址：www.lnkj.cn/uri.sh/8442